"Beautiful writing about Prince Willi
. . . a story about a place that breaks
—Janisse Ray, author of *Ecology of a Cracker Childhood*

"Part celebration, part quest, part hard-hitting critical debate, *The Heart of the Sound* floats us into wondrous new territory. In this graceful meditation about love, loss, and learning, Holleman's prose shimmers like sunlight on clear water."—Nancy Lord, author of *Rock, Water, Wild: An Alaskan Life*

"A love song to one of earth's loveliest and most threatened places intertwined with human loves, disappointments, and healing. An extraordinary book full of wonder and passion and a courageous story about the heart of Prince William Sound and about sounding the depths of our own hearts."
—David W. Orr, author of *Down to the Wire: Confronting Climate Collapse*

"In this vivid and honest account of love, loss, and allegiance to place, Marybeth Holleman engages troubling questions about human impact on the wild. Ultimately, the ravaged landscapes of Alaska and of her own heart teach her to find joy in impermanence."—Lorraine Anderson, editor of *Sisters of the Earth: Women's Prose and Poetry about Nature*

BISON
BOOKS

Marybeth Holleman

the heart of the sound

AN ALASKAN PARADISE FOUND AND NEARLY LOST

University of Nebraska Press
Lincoln and London

Manufactured in the United States of America
♾
First Nebraska paperback printing: 2010

Reprinted by arrangement with the University of Utah Press, University of Utah, Salt Lake City, Utah.

Interior photographs © Marybeth Holleman, except photograph on page xvi © Dean Rand.

Excerpt from "The Peace of Wild Things" in *The Selected Poems of Wendell Berry*. © 1999 by Wendell Berry. Reprinted by permission of Counterpoint.

"Credo" in *The Collected Poetry of Robinson Jeffers, Volume 1, 1920–1928*, edited by Tim Hunt. © by The Jeffers Literary Properties. Used with permission of Stanford University Press, www.sup.org.

Quotes from Matthew Fox and David Orr appear in their interviews in *Listening to the Land* by Derrick Jensen, Sierra Club Books, 1995.

Several of the essays in *The Heart of the Sound* have appeared in different forms in the following journals and anthologies: *Weber Studies, We Alaskans, American Nature Writing 1996, Solo: On Her Own Adventure, The Seashore Reader, American Nature Writing 2000,* and *Orion*.

Library of Congress Cataloging-in-Publication Data
Holleman, Marybeth.
The heart of the sound: an Alaskan paradise found and nearly lost / Marybeth Holleman.
p. cm.
Originally published: Salt Lake City: University of Utah Press, c2004.
ISBN 978-0-8032-3035-4 (paper: alk. paper)
1. Prince William Sound Region (Alaska)—Description and travel. 2. Natural history—Alaska—Prince William Sound Region. 3. Holleman, Marybeth. 4. Prince William Sound Region (Alaska)—Biography. 5. Prince William Sound Region (Alaska)—Environmental conditions. 6. Exxon Valdez (Ship) 7. Oil spills—Alaska—Prince William Sound Region—History—20th century. I. Title.
F912.P97H65 2010
917.98'3—dc22 2010026531

For my son, James
For Prince William Sound
For their future

contents

PART III

When despair for the world grows in me,
and I awake in the night at the least sound
in fear of what my life and my children's lives may be,
I go and lie down where the wood drake
rests in his beauty on the water, and the great heron feeds.
I come into the peace of wild things
who do not tax their lives with forethought
of grief. I come into the presence of still water.

WENDELL BERRY, "The Peace of Wild Things"

prologue

I PULL HARD on the blue nylon line, but it's wound tightly around a jumble of logs half buried in beach gravel. I sit up and lean back against the solid heels of my rubber boots, my back and arms aching.

It's low tide; the cobbled beach slopes down and away, sliding under clear water where mottled pink sun stars cling to barnacled rocks and a multitude of lion's mane jellyfish pulsate with the tide. The water's surface is stippled by rain, and clouds hang low, an embracing shroud that hides the view beyond this beach—the waters of Port Wells sweeping in all directions, to the curving shoreline of Esther Island, of Pigot Bay, of Bettles Bay, and above it all the Chugach Mountains' jagged peaks rimmed with ice—a view I know by heart.

Colors deepen under cover of clouds. Every tree and plant glistens with the lifeblood of this place, a luminous wetness so complete the very rain forest seems to vibrate. Yellow spikes of skunk cabbage jut above emerald sphagnum moss. False hellebore leaves unfurl like waves, downy and deeply ribbed. Tiny wood violets—white, yellow, purple—flash color among pale-green lichen and dark crowberry. Rain drips from hemlock and spruce, their boughs tipped with neon-green buds swelling with the year's new growth. The musky scent of wet earth mixes with the smells of sea and shoreline, and the pungent odor of things washed up.

I bend again to my work, freeing one line only to find another attached to it, along with a ragged piece of fishing net. I call my son,

James, over to use his pocketknife, and together we cut and unravel the knot of lines and net.

As I tug on lines that score my skin, two decades of memories of Prince William Sound disentangle in my mind: the brilliant strands of my first encounters, the frayed webbing of those spill years when everything seemed tainted with oil, the complex knots of years beyond that—and all the changes I never would have imagined. Changes to the Sound that have created a need for this beach cleanup, changes in my own life that have me here, with a son, working for this place, making this small gesture of atonement.

On this last day of May, we are collecting the flotsam and jetsam of human life. It's the second annual Prince William Sound beach cleanup, a fleet of volunteer boat owners and cleanup crews fanning out from Valdez and Whittier on a daunting task. Seven adults and two kids on our boat alone, we spend all day collecting trash from only eight beaches, reaching one side of Cochrane Bay and the northern tip of Culross Island. We fill nearly one hundred garbage bags, most of it derelict fishing gear washed up in the storms of winter.

Some of the detritus we recycle. Michelle takes some yellow line for her dog; Mike takes a piece of netting for his neighbor's peas to climb; Elijah, six, has a bag filled to the brim with buoys, driftwood, rocks, and shells; and James, at eleven, still finds treasure he deems worth keeping: a weathered wooden oar, an orange boat bumper, a padded swivel seat from a boat.

There are questions I didn't even know to ask when I first came to Prince William Sound. These are questions that now feel increasingly vital to living here, to living anywhere on Earth. These are the questions that are entwined in my own life, that bind this story of my first fifteen years in Alaska with the Sound.

How do you form a relationship with a place? What forces, both inherent to you and inherent in the land itself, make that relationship happen? How does it affect the place? How does it transform you, inform your life, feel in your body? What do you do about it, once you've fallen in love with a place? How does this love make you act?

I have come to know what we ask of the places where we dwell.

We ask for all that sustains our lives and all that gives life meaning. We ask for security and passion, for solitude and community, for sanctuary and joy. Prince William Sound has taught me that an abiding relationship with a place can encompass all the challenges of an enduring relationship with another person, all the challenges of acceptance, compassion, reciprocity, growth, and, finally, unconditional love.

On our return to Whittier, the rain pours from the sky as if it will never stop, not even after forty days and forty nights have passed. I slide the boat's side window open enough to put my head out, feeling the spray of wind and rain on my face. On the water's rippling surface, two marbled murrelets float like little brown buoys. A sliver of fish dangles from one murrelet's beak. The other dives. Moments later, a luminescent circle of light appears, a cluster of bubbles erupts, and hundreds of small fish burst into the air. In the center of the dome of leaping fish is the marbled murrelet. In its mouth, a single thread of silver.

CHUGACH
Port Wells
Passage Canal
Whittier
Whittier Icefield
Perry I.
Naked I.
Port Nellie Juan
Dangerous Passage
Knight I.
Sargent Icefield
Chenega Bay
Montague I.
5 0 5 10 15 Miles

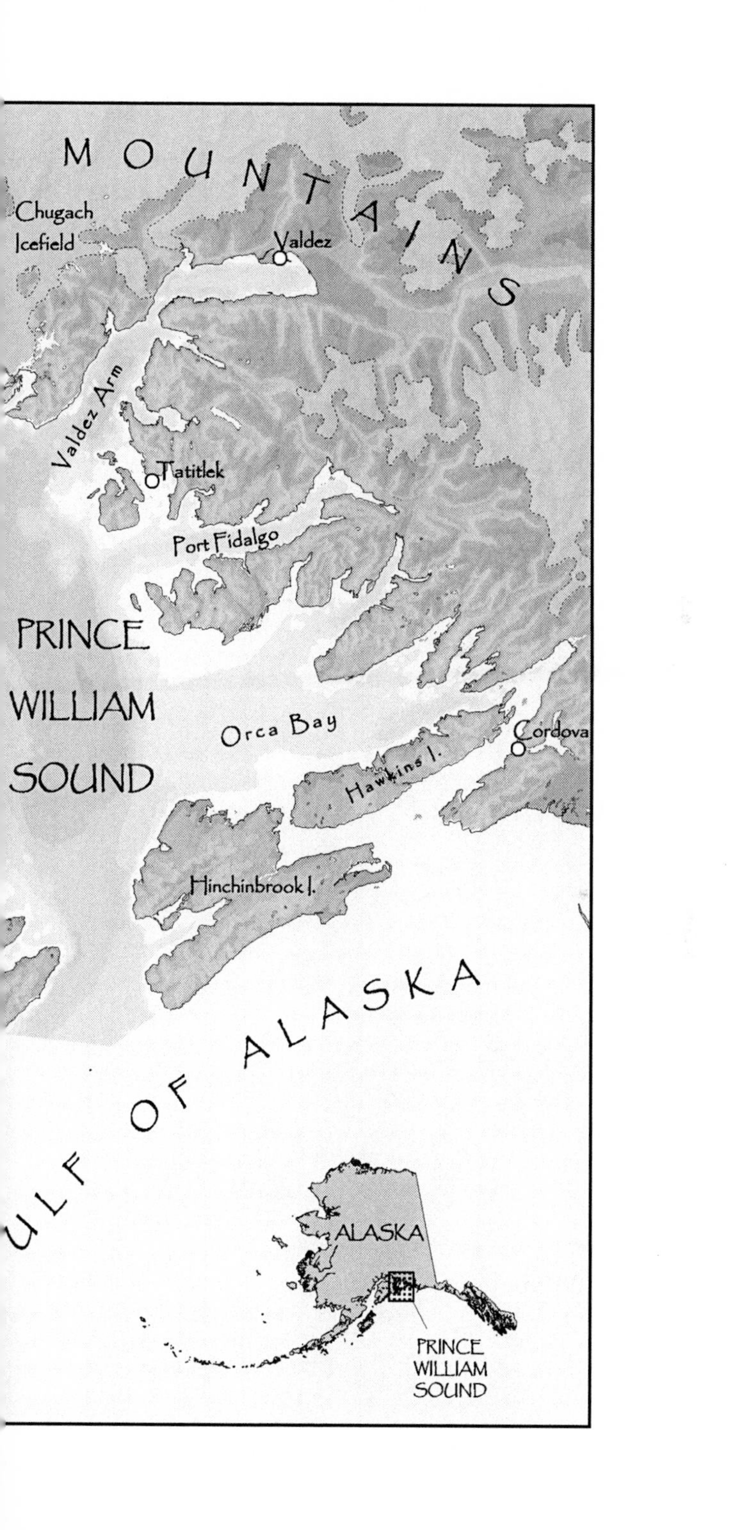
M O U N T A I N S
Chugach
Icefield
Valdez
Valdez Arm
Tatitlek
Port Fidalgo
PRINCE
WILLIAM
SOUND
Orca Bay
Cordova
Hawkins I.
Hinchinbrook I.
ULF OF ALASKA
ALASKA
PRINCE
WILLIAM
SOUND

the heart of the sound

PART I

We have to fall in love. And we have to de-anthropocentrize this thing about falling in love, which our culture reduces to soap operas and finding a mate. We have to realize you fall in love with creation.

MATTHEW FOX, in *Listening to the Land*

parts unknown

1986

That first summer, I sold tickets on the Whittier Shuttle. This train running between Portage, on the Alaska Highway, and Whittier, on the shores of Prince William Sound, was the only land route to the western Sound.

It wasn't the type of adventure I'd had in mind when I answered the magazine ad for summer jobs in Denali National Park. Before my husband, Andy, and I arrived, we looked up the town of Portage in the *Milepost,* the mile-by-mile guidebook for traveling the Alaska Highway. All it said was that Portage had been destroyed by the 1964 Alaskan earthquake, leaving me to imagine some desolate little spot with crumbling houses and dead trees. We were chosen for this job because of the earthquake—there was no longer anywhere to live in Portage, and we carried our home on the back of our truck. We set our truck camper beside the tracks and rode the train back and forth every day.

The train stop in Portage was nothing more than a gravel parking lot sandwiched between the highway and the tracks. A few houses sprawled, sinking into tidal marshes, their rooftops bowed by time, old trucks and bedsprings strewn about. Around them stood long-dead spruce trees with limbs and trunks bleached as bones. But these ruins lay at the head of Turnagain Arm, whose granite-colored waters rimmed by snow-topped mountains stretched westward. And in the other direction, just beyond the tidal flats and through two tunnels, was Prince William Sound.

As soon as I had walked the aisles selling tickets, and the train had eased away from the Portage train stop, I retreated to the baggage car—the place I liked to ride. Surrounded by dry bags, backpacks, coolers, nets, fishing poles, bags of groceries, and rows of rubber boots, I slid open the heavy metal doors and leaned against the door frame. There was nothing between me and this new place, and the beauty of it all rolled by me.

The path of the train led across tidal flats awash in purple lupine and swaying grasses. Arctic terns stroked the air, hovered, and dove into clear ponds after silver fish. Pintails, mallards, and widgeons glided among yellow lilies.

Beyond these flats, the train ran between Portage Creek and the Chugach Mountains. Portage Creek was swift and narrow, its waters a milky blue from Portage Glacier. The sheer sides of the mountains were streaked with waterfalls and studded with cliffs, deep forests, and tundra.

Right before the first tunnel, a section of tracks was askew enough that the familiar clacking of the train's wheels grew louder. Behind my baggage car, the cars, trucks, boats, and buses on the flatbed cars began swaying in rhythm with the sound of the train. In front of the baggage car were the three passenger cars that carried foot passengers, who sat on cracked green leather seats and looked through permanently fogged windows. Beyond these cars, the gleaming yellow-and-blue Alaska Railroad engine churned down the tracks.

In my first letter home to my mother, I told her she'd love Alaska for two reasons: one, there were more waterfalls in Portage Valley and Prince William Sound than I'd ever seen; and two, they let dogs ride the train. Not in the baggage car in kennels, but in the passenger cars, up on the seats, right beside their people—beside fishermen just back from an opener smelling of fish guts, neatly dressed elderly tourists just embarked from their cruise ships, college students with backpacks ready for a summer of adventure, Anchorage residents heading out on their sailboats.

Entering the first tunnel, the train was engulfed by absolute darkness. The air was moist and cool, and the engine's roar deepened. One mile, and it emerged into Bear Valley. Portage Lake, jammed at this lower end by house-sized icebergs from the quickly retreating Portage Glacier, spread south on one side. On the other side, an open birch and cottonwood forest was framed by mountains crested with snow. Sometimes the train would slow almost to a stop, and a couple of backpackers would jump off to hike for a few days in Bear Valley. Once, as the train emerged from the mountain, a black bear scrambled up the mountainside right beside the train, startled by the sudden appearance of this rumbling beast.

On the other side of Bear Valley, the train disappeared into the two-and-a-half-mile-long tunnel. Absolute blackness again, only the engine light showing the damp jagged rock inches from the track. The engineer always carried a flashlight because, he told me, "Can you imagine what kind of mayhem there'd be if this engine were to die in the middle of this tunnel? There would be no light, none."

Built quickly during World War II, this tunnel was barely tall and wide enough for the train. I'd heard that hikers sometimes tried to go out through the tunnel; one got caught by a freight run, and had to press tightly against an indentation in the rock to avoid being struck.

On one trip, the engineer let Andy and me ride out on the nose of the engine, nothing but a thin chain between us and the track. For one intense minute, the engineer cut the headlights. With the complete absence of sight, the rush of smells and noise and blackness was amplified. I screamed, as much from joy as from terror. No one, not even Andy swaying by my side, could hear me.

After straining in the darkness for a dim spot of light, the train burst into light again, and Passage Canal lay before us. Every time the train emerged into that light, I felt as if I were being reborn. After the darkness, to be bathed in light, to see mountains awash in green forests and crowned with glaciers, and water spreading out farther than I could see.

Prince William Sound. I had never heard of it before arriving in Alaska. Now I stood at the northwestern edge of this vast and intricate web of ice fields, mountains, forests, and sea, yearning to be out in it.

The train conductor, Dan, a brash and warmhearted Irishman, was after us from the beginning to find some way onto the waters of the Sound. With sheer mountainsides dropping straight to the sea on either side of Whittier, water was the only passage.

"C'mon," he'd say in a thick accent amplified by years of working around trains, "you're not telling me you came all the way up here to spend your time on this train."

He offered to let us borrow his old fishing boat, but it was weeks before we took him up on it. Andy had much more allegiance to the railroad and our jobs than I did. Since our subcontractor hadn't given us instructions about what hours we should work, Andy figured we should work all the train shifts. I had come to Alaska for the adventure; the job was just the entrance fee. I couldn't bear standing on the threshold of such a magnificent place and not venturing out into it.

Finally, though, we agreed to take a day off, and arranged with Dan to use his boat. We stopped first at the Hobo Bay Trading Company. Babs, owner of this burger joint and a longtime Whittierite, gave railroad employees a daily discount of one dollar. Usually we spent it on a couple of her jumbo-sized brownies, but this day we got the works: two burgers, fries, drinks, and the brownies.

Dan's boat looked like a bachelor pad; it had a musty odor, and every surface was coated in grime. But it floated, and it was free.

We climbed aboard and putted slowly out of the harbor. When Andy gave it more gas to try to get the boat's bow to plane out on the water, the engine overheated and died. The boat had sat idle too long. We drifted around, Andy fiddled with the engine, trying a few more times to get it running smoothly, then we limped back to the harbor.

But Dan wouldn't give up his crusade. He cut a deal with the owners of Whittier's only kayak rental service: free train rides for them and their van in exchange for free kayak use for Andy and me. Within a week, we were out on the water again.

Lois, who with her husband, Perry, owned Prince William Sound Kayak Center, picked out a double kayak for us, along with a bewildering medley of gear: spray skirts, rubber boots, hatch covers, and double paddles. Andy and Lois and I carried the kayak down to the water's edge next to the ferry dock in Whittier.

Whittier was not picturesque. Built during World War II as a navy supply port, its site had been strategically chosen: frequently socked in with rain and fog, it was hidden from enemy planes. The tunnels were blasted out in record time—the kind of construction project that could only have been accomplished through the dogged commitment of a country at war.

The remnants of that wartime base still dominated the place. A tank farm consisting of a half-dozen giant white cylinders crouched next to the tunnel entrance. Closer to the harbor and train stop, two large buildings loomed over all: one, formerly the barracks, was a gutted concrete shell; the other, fifteen-story Begich Towers—named after a state senator who lost his life in a plane crash on Portage Glacier—housed all the residents, the school, health clinic, post office, and an array of small stores. A fisherwoman from Cordova once told me, "Begich Towers reminds me of a mental institution I once worked in."

In front of these buildings was the train yard. A web of tracks dominated the townscape of Whittier, many of them lined with empty freight cars awaiting container barges from Seattle. A few railroad warehouses were scattered among these tracks, but there was no train station. This port was much more important for carrying barge freight to Anchorage than for carrying passengers to Whittier.

Beyond the train yard were the docks and a small boat harbor. A few scattered buildings lined the shores, among them the harbormaster's office, Babs's place, and a couple of storage units. The small boat harbor was filled with an assortment of sailboats, yachts, small powerboats, commercial fishing boats, and tenders. One road curved from Begich Towers through the train yard to the harbor; there were no single family houses, no neighborhoods, no parks, no downtown.

Though several thousand cruise ship passengers were funneled through here every summer, the only amenities for them were an old railroad-car visitor center and a cluster of port-a-potties between the cruise ship dock and the train stop. There was nothing for tourists to see in Whittier, nothing for them to do. For many in the tourism trade, Whittier made good fodder for insider jokes. One bus driver who brought tourists from Anchorage to Whittier to board cruise ships told us that when his passengers asked him, "What's Whittier like?" he'd answer with a simple question, a joke they didn't get until their train arrived in Whittier: "Have you ever been to Paris?"

It was an odd smattering of civilization, this town, a strange gray pockmark at the end of verdant Passage Canal. It was a good place to be leaving.

In her buoyant voice, Lois gave us a twenty-minute lesson in kayaking.

"Nothing to it," she said, as we made our first attempt at getting our feet and legs into the narrow openings without tipping the kayak. She showed us how to put on our spray skirts and ease our bodies into the narrow opening of the kayak, me in front and Andy in back. She gave us instructions on paddling, using the rudder, and getting in and out.

Then we were on our own.

We dipped our paddles into the water and pulled easily, effortlessly, away from the shores of Whittier and into sun and flat waters. After paddling steadily for a while, I shed layers down to a tank top. I had worked the train enough to know that this warmth was unusual in the Sound. Many days we would be in sunshine in Portage and see the train returning from Whittier gleaming wet, see the passengers step off in soaked rain gear and boots. The Sound is a temperate rain forest, absorbing several hundred inches of precipitation a year. A sunny day like this was a rare gift. The kayak was light and maneuverable, and we glided across Passage Canal in no time at all.

6

We aimed the kayak's nose to a kittiwake rookery and waterfalls on the opposite side of Passage Canal. We could see the falls from Whittier, but not the kittiwakes. Almost two thousand black-legged kittiwakes nested on the rock cliffs around the falls every summer; it was, I'd been told, one of the largest black-legged kittiwake rookeries in Prince William Sound.

As we approached, an occasional cry of a gull became a constant cacophony. Threading through their cries, rising like their wing beats overhead, was the roar of the waterfall rushing down granite into the sea. This whirlpool of sound and motion drew us in.

At the waterfall and rookery, we stopped paddling to take in the sheer face of the three-hundred-foot cliff, where hundreds of white birds flew and perched on the narrowest of ledges. Where the largest concentration of birds roosted, only a few green plants clung to vertical rock. On the margins of the rookery, where the concentration of nests thinned out, every roothold was taken. Bluebells, orange columbine, bracken fern, and saxifrage clutched the rock face with tenacity that belied their fragile appearance.

The kittiwakes were white, with gray backs and black legs and wing tips. Small graceful flyers, they resembled arctic terns more than the glaucous-winged and herring gulls we usually saw. Having returned to this rookery in May, they would leave when their young were fledged in late August; the rest of the year they lived and fished on the open ocean.

In some places, fuzzy gray chicks huddled against the inexorable pull of gravity on a shelf only a few inches wide. A parent brought fish to a chick, the chick becoming a gaping beak to receive it. Some alarmed parents began dive-bombing us, swooping low and fast, pulling up just before their sharp beaks hit our heads—clear warning that we were too close.

We let the outgoing tide carry us along the shore, our paddles resting across the kayak. Mist from the largest waterfall coated my upturned face. Dense emerald moss punctuated with maidenhair fern covered the rock along the waterfall's border.

The thunder of water falling and the high cries of kittiwakes slowly receded. Another sound grew more distinct, a lighter, higher water sound; a smaller waterfall came into view. Drifting by, a slight mist blessed me again: near silence. Only small waves lapped the pebbled shoreline. I let my head drop back, warmed by sun and soothed by quiet.

©

My grandfather Jiulio Bruno came to the United States from Montaguto, Italy, when he was twenty-three years old. He traveled through Canada, taking a backdoor route that included working for German cheese makers near Toronto. He settled in Youngstown, Ohio, where several of his brothers had moved before him. As a child, I had been entranced by his coming-to-America story, by the courage and adventurous spirit of this young man seeking his fortune in a new land where he didn't even speak the language.

When Andy and I married and bought ten acres of land outside Chapel Hill, North Carolina, my grandfather was more thrilled than anyone in my family. He and my grandmother—Vera Iagulli Bruno, who had also emigrated from Montaguto, through Ellis Island, at four years old with her entire family—had nurtured their three acres with all the attention of family farmers. On every summer visit to their home, all through childhood and into adulthood, one of the first things I'd do upon arrival was to run out back, to their cold frames bursting with the sharp scents of tomato and pepper seedlings, the rows of corn and sweet pepper parading off toward the woods, the gnarled and winding branches of muscadine grapes bowed by globes of fruit. Inhaling the fertile exuberance, I felt at home.

On our new land, Andy and I planted several large garden beds, a cornfield, a few apple and peach trees. I made an herb garden outside the kitchen door, and Andy turned our compost pile into dark, rich soil. Grandpa gave me seeds from their oxheart tomatoes and banana peppers; he gave me a shoot off the fig tree he'd started from the one on his family's farm in Montaguto.

When we bought the truck camper, quit our jobs, and began traveling the country, he called us foolish, called us gypsies. When we ended up in Alaska, he was even more disappointed. It was as if I had jumped off the rim of the earth. "Alask'!" he'd say to me. "Why Alask'? Too cold to grow anything!"

He couldn't fathom why anyone would choose to leave a dependable job and a good piece of land. I might have viewed myself as off seeking my own fortune in a new place, emigrating across the country the way he had once emigrated across the Atlantic Ocean. But his journey was driven by economic necessity, and once he had land and work, he never wandered again. He always spoke longingly of Italy, "the Old Country," sometimes dreaming aloud of going back there one day. Would I do the same someday for North Carolina?

I knew the experiences of our generations, and our lives, were very different; I knew that the Italian sense of rootedness in land ran deep; I knew that an Italian's emotions, both good and bad, were sharply chiseled onto the surface of a life. All this made his disappointment in me only harder to bear. I had reveled in his approval of our land purchase, of our attempts at gardening. To lose that was a deep blow. If I could have shown him Alaska, shown him the glistening waters of Prince William Sound, the opportunities for a life fully lived that I sensed in this place, would he have understood? If I could have told him about Alaska's homestead lottery, a state program that offered free land—*free land!*—would he have approved of my move? Had his family in Italy shaken their heads at his departure to the New World? Had they wondered how he could possibly take root so far north of the sweet Mediterranean sun?

☙

We drifted onto a wide curve of tidepool-studded beach with a forest of dead spruce trees as its backdrop. The 1964 earthquake, its epicenter about ten miles east, had tilted the entire Sound. Here in the northwestern Sound the land sunk up to twelve feet, while on Montague Island seventy miles to the southwest it rose forty feet. High tide brought in saltwater to tree roots, mummifying them. These trees still stood, needles and bark long gone, trunks burnished silver by decades of storms.

Behind them, on a slight rise, was a dense forest of still-living spruce and hemlock; their verdant limbs were draped in clouds of old-man's beard. A freshwater stream, filled with silty blue meltwater from a hanging glacier above, clattered over fist-sized rocks to saltwater.

All down the length of Passage Canal, mountains rose straight up from the water so steeply that there were only a few places, like this one, where we could beach our kayak and walk the land. Where they brushed the edge of the sea, the mountains were covered in a deep green forest canopied with spruce, cedar, and hemlock. At the tree-line, only one thousand feet up, low-growing shrubs and herbs took hold; above them were rock and ice. Ice fields and glaciers, tongues of ancient snow alchemized into blazing blue ice, hugged sharp peaks and poured down valleys. Gushing from glaciers, streaking from the ice fields above, threading through granite crevices and spraying over mossy beds were dozens of waterfalls plunging into the fjord.

The iridescent blue-green waters of Passage Canal stretched fourteen miles to the southeast. There, they broadened into Port Wells, and then into the entire Sound, a ten-thousand-square-mile body of water that spread beyond my view like the white space labeled "Parts Unknown" on an early explorer's map.

Pink against gray, green strands in a sinuous curve, clusters of brown and black, radiating fingers of red and purple formed a collage of color and texture beneath the tidepool's glassy surface.

Crouching, I put my hands on solid granite surrounding the pool and rocked forward slowly, letting my eyes adjust to this new spectrum as though I were moving from daylight into dark. A darting movement exposed a small fish mottled brown, tan, and black to blend with the granite pebbles and bits of shell layering the bottom of the tidepool. A dark-gray shell lifted; sprouted delicate, articulated legs; scurried off.

I leaned in. More shells dashed about, tentacles of red and green rippled, purple and orange legs stippled with white emerged from the rock, a pink mat covered gray rock, all of it entwined in bright-green cellophane leaves, grass-thin blades, and burgundy fronds.

I bent lower, my face almost touching the water's surface, trying to pick out each individual, its particular movement and shape and hue. I wanted to learn their names, their lives, the inhabitants of this small tidepool.

This beach, this tidepool, was only a few miles from Whittier. But it was far across the water, barely visible, and only a few boats had passed by all day, their motors a brief, distant hum. There was no one else around, no sounds except those of water and wind and the keening of birds. I felt as if I had traveled far in this one afternoon, as if I had emigrated across an ocean.

Like the center of a wheel, Prince William Sound is the apex of Alaska's Pacific shores that curves from southeast Alaska to the Aleutian Islands. It begins after the six-hundred-mile stretch of coastline between Glacier Bay at the northern end of Alaska's panhandle to Cape Suckling, a broad beach ravaged by Gulf of Alaska storms that is called "the Lost Coast" because its land and waters are so untraveled by humans.

Just west of Cape Suckling lies Kayak Island, first landfall of Europeans on Alaskan soil in 1741, and the first welcome—or warning—of the convoluted coastline to come. Beyond Kayak Island and the Copper River Delta, the largest wetland on the Pacific Coast, the shoreline bends northward and begins a thirty-five-hundred-mile-long undulation of capes and fjords and islands and islets and sea stacks. On the mainland, three mountain ranges—the Chugach, St. Elias, and Kenai—encircle these waters with peaks reaching fourteen hundred feet. These mountains are draped in ice fields from which more than one hundred and fifty glaciers pour, nearly twenty of them scouring rock down to the tide, where their ancient ice cracks and calves into saltwater. Montague and Hinchinbrook, the Sound's largest islands, spread like sentries between the Sound and the Gulf of Alaska, tempering the fury of Gulf storms.

Alaska's coastline, altogether as long as the coastline of the rest of the country, is deeply indented for much of its length. But Prince William Sound, with thirty-four major islands and hundreds of smaller ones, and more than one hundred fjords, bays, and passages, is by far the most ragged. It is an expansive, meandering maze.

I liked that. I liked reading the map of the Sound and tracing the fingers of water yet to explore, the beaches and ridges to walk, the headlands and islands to stand upon.

Early one morning, selling tickets at the Portage train stop, I had met a man who had fished the waters of the Sound for twenty-three years. Tall and lean, with a long black beard and a black cap, he'd told me story after story in a deep Russian accent about his adventures in the Sound.

"I could spend many lifetimes exploring Prince William Sound," he had said, "and still not see it all."

When I was eight, nine, ten, I spent hours on the roof of my parents' garage, where Dad kept the things he still wanted to fix. Pine and mimosa and white oak surrounded me, kept me out of sight of my brothers and sisters playing in the yard below. Sometimes I'd hear the rhythmic sound of my brother Joe dribbling and shooting the basketball. I almost always had books with me.

Once, I had the "A" book of our family *World Book Encyclopedia*. I'd finally found the place I wanted to live when I grew up: Argentina. It said right there in the *World Book* that you could travel from the mountains to the coast in a half hour. I could see from the map how close they were, steep, snowcapped mountains and sunny beaches. That's what I wanted. I wanted mountains like the Appalachians I lived in now, but I wanted ocean, too. Summer trips to Myrtle Beach with my family were never enough. I would keep the discovery of Argentina to myself. My family would just tell me that it's jobs and families that decided where you lived, not mountains and oceans.

Somehow I missed finding Alaska in that "A" book. But now I'd found Prince William Sound, where mountains and sea were wedded along every shore.

On that rooftop, in those Appalachian Mountains, that coastline of my childhood, I had known moments where I'd felt the whole of the place and me in it, a dissolving of the wall between me and everything else. A deep sense of connection. Among the blueberry meadows atop Mount Pisgah. Flowing down Sliding Rock like another drop of water. Perched on a sand dune on the Outer Banks. Climbing Chimney Rock in the fog.

How surprising to have found that feeling here, so sudden and sustained.

@

At the sound of popping, I looked toward the tide line. A tangle of seaweed in the low-tide zone was being reclaimed by the rising tide. I walked to the edge of the water and sat on my heels. Long green strands of eelgrass sprawled before me, tossed and tangled like uncombed hair across a pillow. A large sharp-edged rock jutted above these strands, its sides coated in barnacles and blue mussels. I crept across the slippery eelgrass and peered around at the underside of the rock. A green anemone clung there, its tentacles tucked into a soft green belly that was veined with red.

It wouldn't be long before it was once more in the sea, and those tentacles would blossom, groping for microscopic plankton. Prince William Sound's tides reached some of the greatest ranges in the world. At a spring tide like this one, on a new moon, the tides could rise and fall up to twenty-four feet. The beach we had landed on at low tide was shrinking, its shape changing every moment.

Hearing the popping sound again, I lifted strands of eelgrass, finding more blue mussels and a twelve-legged sun star the color of fresh apricot. Then the rising tide reached the eelgrass; the long strands lifted, each blade pulling free and swaying, creating an underwater grassland at my feet.

I placed my hands on the eelgrass and felt the incoming rush of water. Cool and silky smooth, the water lapped at my fingertips, threaded between fingers and up over each hand, embracing them with the pulse of the sea.

I did not want.

Duck

Culross Island

Beach the boat on Culross Island and strike out into the woods. No trail, just a point on the map marked "Duck," the highest point on the island. Head up. Thrash through dense Sitka spruce and hemlock forest, hop across muskeg meadows spongy with sphagnum. In and out of forest cover, watch all the while for bears and river otters. Gaining treeline, clamber through a thicket of alder and willow; climb up alpine meadows glowing with mountain heather and crowberry, onto a rock spine coated orange, yellow, and green with lichen; follow it to the peak.

Look over the other side, see Hidden Bay, a string of oval bays silvered by the evening sun. Beyond are Perry, Lone, Naked Islands, and farther still the snow-tinged Chugach Mountains of the mainland.

Stretch out arms, gather mountains, bays, islands in a wide embrace. Call it love.

locus amatus

1986–1988

PETER WAS TOWING our kayaks with his fifteen-foot Zodiak. We traveled slowly, worried that a wake might roll them. We stopped often to adjust the lines. All the way out Passage Canal, I faced backward instead of forward to keep an eye on the boats, robbed of the chance to scan the water for sea otters or the shore for bears.

I trusted Peter. As brakeman, he had one of the grimiest jobs on the train, and his home was nothing more than a wall tent between the tracks and the shoreline in Whittier. He was young, quiet, a loner. But I had been inside the wall tent; there was one cot with a sleeping bag, and a cardboard box for a table. On the box he kept a daily weather log, the calendar lines drawn neatly, the writing like that of an architect. His fastidiousness extended to his boat. He kept it in a heated storage unit in town. When he carried it out for this trip, he first turned it over, using a trouble light to check for any small holes in the hull. The engine shined like new, though it was several years old.

Peter was taking Andy and me and two single kayaks to the mouth of Passage Canal so that we could explore Blackstone Bay to the west. We planned to have him carry us to Decision Point, twelve miles out. We'd camp there for three nights, then paddle back to Whittier.

What I wanted was to paddle the fourteen-mile length of Blackstone, past Willard Island to the face of two tidewater glaciers, Blackstone and Beloit. I had yet to see a tidewater glacier; the only glacier

I'd seen up close, in Wyoming's Teton Range, was no more than a year-round snowfield in a bowl at the mountain's base. What I wanted to see on this trip was a true glacier's towering, ancient ice right in front of me; I wanted to drift among icebergs.

While waiting for Peter at the Whittier dock, I had talked to two kayakers who had just returned from a long, wet trip. Wet sleeping bags, tent, raincoats, pants, and maps spread around them in the clear morning light, the first sunshine in twelve days. I asked them where they had paddled. In reply, the man picked up a soggy USGS topographic map of western Prince William Sound and showed me their route.

They had paddled for two weeks up the entire western side, at least seventy-five miles around islands and across straights, past glaciers and through passages—a route that I hadn't even considered possible by kayak. I traced on the map the intricate maze of land and water around Icy Bay and Dangerous Passage—the sinuous blue scribing waterways, the contour lines delineating spiring peaks, the white bands promising tidewater glaciers. My pulse quickened as I imagined it all, and me in it.

"That looks beautiful," I said. "I'd like to go there someday."

"Well, here, take these maps," the woman said to me. "We don't need them anymore."

I thanked them, tucked the wet maps into my coat pocket, and walked back to our kayaks, feeling the immensity of this place stretch out before me.

Now, the three of us motored down Passage Canal slowly. In the three hours that leg of the trip took, clear skies dissipated once more into rain and wind. The clear image I held of Blackstone Glacier's icy face thrust against a blue sky became as blurred in my mind as the view through my rain-streaked glasses.

At Decision Point, Peter maneuvered onto a protected beach on the back side of the point, and we unloaded. The small island of forest looked untrammeled from afar, but up close I could see a couple of round, flattened areas under trees, a fire ring filled with stones burned pink, and a half-burned log.

All around was wet. I searched for firewood, but every piece of wood was heavy with rainwater. Then I remembered the railroad flares.

Before we left, Dan, the train conductor, had handed us a shoe box filled with foot-long red tubes.

"Take these," he'd said. "You'll need them for starting fires."

Andy laughed. "That's OK. I know how to build a fire."

"No, you take them," Dan insisted, his ruddy face inches from Andy's. "You just don't know what it's like yet out there. One sunny day like this isn't going to dry anything enough for a fire. Take them. They'll burn long enough to get some wood going."

"Thanks," I said quickly.

After Dan walked off, I told Andy, "We can always use them for emergency flares."

As it turned out, this was a five-railroad-flare night. All the wood was rain-logged, saturated to the core. Just as the flares began to dry some wood, just as my hands began to get warm, the rain stopped. Within minutes, hordes of no-see-ums appeared, flying in my eyes and nose and mouth. We fished out the head nets.

I had never seen such sodden wood, or so many small bugs swarming. I had never been so cold and wet in the middle of summer. But this was a rain forest, the northernmost reach of a temperate rain forest that extends down the Pacific Coast to northern California. It couldn't be a rain forest without rain. I had seen the buttons at the Whittier visitor center: "Whittier Rain Festival: 365 days a year."

The rain, the soggy wood, the bugs, and the cold made everything a struggle. But we managed to put up the tent, fix and eat dinner, wash up, hang the food bag out of a bear's reach, and crawl into the tent. My sleeping bag was clammy, and I awoke countless times in the night to the patter of rain on the tent.

Water shapes Prince William Sound, water sustains it, water creates and contains. This is what I was beginning to learn. This is what I came to love.

Water shapes Prince William Sound with long arms of ice, glaciers pushing and carving out deep valleys and fjords. All the valleys and fjords are U-shaped, rather than the V-shape of older mountains or mountains formed by uplift, like the Appalachians of my childhood. It is all so new here—the last ice age covered the entire Sound only ten thousand years ago, and erosion hasn't yet had time to break down rock, to diminish the angles of mountain slopes and spill soil down toward the valley's center. The ice still covers a third of the Sound's land. Beneath the waves, the steep slopes continue; the bottom of the fjords has that same deep-cut U-shape.

All this is visible. At Blackstone Bay, the water becomes suddenly shallow just offshore of Willard Island. Just underwater is a band of rock rubble left by the receding glaciers in the fourteenth century, a terminal moraine that stretches across the entire bay. On one trip to Nellie Juan Lagoon, I walked up beside the glacier and ran my fingers over deep scratches in rock made by the glacier, runnels that a giant out of Grimms's fairy tales might have made, digging his nails in the solid granite and clawing his way up.

It wasn't long after the glacier's retreat before plants took hold, because Prince William Sound is also sustained by water falling from the sky as rain or snow into cracks and crannies of that glacially striated rock. Water streams out of the forest, thousands of waterways rushing to the sea, softening the sharp glacial-cut mountains, nourishing rooted life on every inch.

For all its fecundity—the way green appears in every viable crevice and ledge—this rain forest is fragile, a thin membrane between sea and rock and ice. Only 15 percent of the Sound's land is forested, and the soil is often no more than a fingernail thick. On this thin soil, it is rain that sustains all life. The old-growth forests, which still constitute much of the Sound's forests, contain trees that are several arm's lengths around. But they grow slowly. A five-hundred-year-old spruce in Prince William Sound is equal in size to a forty-year-old pine in North Carolina.

Water links land and sea. The Sound is defined as a fjord-estuary because so much freshwater, from streams and from glaciers, flows into it, carrying nutrients into the sea. The sea delivers nutrients back into the forest as well. Wild salmon migrate by the tens of millions to more than two thousand streams to spawn and die. Seabirds flock to the Sound's forests and cliffs each summer to raise their young. Harbor seals and Steller sea lions haul out on land to rest, mate, and raise young. On land, Sitka black-tailed deer, brown and black bears, river otters, and weasels all rely on the edge of the sea for sustenance.

Prince William Sound is a fluid world, a mirror of the earth itself. The surface of the earth is, I knew, 71 percent water. Just like my very body, the world around me was mostly water. Two hydrogen and one oxygen, molecules in concert, a fluidity around and within that defies the solid life, undercuts the unchangeable with motion, washing away and re-creating, over and over.

©

On our first visit home, Andy and I showed slides of our trips in Alaska, in Prince William Sound, to friends in Chapel Hill. They saw images of harbor seals on ice in front of Blackstone Glacier, of the tapestry of islands and islets in Culross Passage, of a bald eagle standing lookout on a rocky spire, of a stream choked with red salmon in Picturesque Cove. They barely said a word. They watched, but asked no questions. I felt as if we were showing home movies to captive dinner guests: what we did on our summer vacation. It was the same response we'd had showing the slides to my family.

My family's lukewarm reaction hadn't surprised me; they rarely traveled, and when they did, it was to New York, or Disney World, or Paris—adventures of culture, not wilderness. I had believed I had more in common with my friends, with whom I had shared so many adventures, and, I thought, a love of wild places. I had expected them to jump up and say, "Oh! I have to go there! It is unlike anything I've ever dreamed!"

But they didn't. Not a one. It was as if this landscape so dramatically different from their own were a foreign language they didn't care to learn. I struggled for words that would illuminate the images, but instead felt a chasm between us and between the places we called home. Instead it was I who, every time, to myself, would say those words, feel that excitement, each image urging me to return.

In those days, it was easy to have a bay, a fjord, a beach all to ourselves; it was common to not see or hear another human for days. There were other people out there, but they were diffused over a place the size of the state of Maryland. About seven thousand residents in five communities clung along the Sound's thirty-five-hundred-mile shoreline. Whittier had several hundred year-round residents, a few of them commercial fishermen. Cordova and Valdez each had a few thousand, but they were more than one hundred miles away on the eastern side of the Sound from where we spent most of our time. The two remaining Native villages, Tatitlek on the eastern Sound and Chenega Bay on the far southwestern corner, had around one hundred people each.

Nearer to Whittier, a few dozen commercial fishing boats crowded certain places like Maine Bay, Eshamy Bay, and Esther Passage during salmon runs, and some fished for bottom fish as well. But they clustered together and were around for only a few weeks or days at certain times of the year, like the birds that migrated through in spring and fall.

I liked the rarity of people. There was a freedom in it, to be in such a remarkably different place from where I grew up, so far from anyone who knew me, from anyone's expectations. A chance to just be, a kind of acceptance I could take for granted.

"This is quite a state we're in," Andy and I would say to each other, laughing. If this place were on the East Coast, we knew, it would be inundated with people in all kinds of boats. Overrun, like Jordan Lake, created by a dam near our home in Chapel Hill: after a few years, it was so crowded with boats that we quit going there. The birdsong had been drowned out by whining outboards, the still waters where heron fed were churned by crisscrossing wakes, and the peaceful reveries were undone by the threat of boat collisions.

We spent the week of July 4, 1988, in Harriman Fjord. We drifted in front of Harriman, Barry, and Surprise Glaciers; we camped on mountainsides between dense forests and spongy meadows; we walked beaches thick with ryegrass near the glaciers. For the entire week, we saw only one pair of kayakers and one tour boat—a double catamaran on a day trip that plowed down Harriman one afternoon, leaving a wake so big it nearly overturned our inflatable anchored onshore. Except for that brief tear, the fabric of the trip was seamless.

When we returned to town, we read in the newspaper that Harriman Fjord was one of the most popular places to be that holiday weekend. We laughed, thinking it a far-fetched conclusion, unless the reporter was referring to something other than humans. It was a popular place that weekend, popular among arctic terns, killer whales, bald eagles, harbor seals, puffins, oystercatchers, and sea otters.

Some folks in Whittier thought we were crazy to go out in a kayak or a small inflatable. They told us the seas could get so rough they'd swamp our boat; they told us the water was so cold that, if we fell in, we'd die of hypothermia after fifteen minutes; they told us that if we did get in trouble it would be too late before a rescue could reach us.

We knew of only two others who had ventured out in inflatables. One was Peter. Another was Tom, wired and wiry, who proudly told me how crazy folks thought he was for long-lining for halibut in a little inflatable. Laughing, black hair tousled and brown eyes wide, he told me, "Lost all my gear overboard in a storm yesterday. Barely escaped with my life."

It was the extremes I found in Prince William Sound that most appealed to me—extremes of weather, landscape, wildlife. One of the most active seismic regions in the world, the Sound anchors the eastern edge of the Ring of Fire that arcs along the Aleutians to the Kamchatka Peninsula. Prince William Sound is the northernmost reach for three evergreens: Sitka spruce, western hemlock, and Alaska yellow cedar. It is a rest stop and destination for tens of thousands of migrating birds and marine mammals, birds like arctic terns who fly from the southern tip of South America just to nest in Alaska's summer light, whales like the humpbacks who swim from Baja California and Hawaii to feed on rich plankton blooms.

It is a place of convergence—the geographic center for Alaska and the Pacific, where the Arctic to the north, Aleutians to the west, and Inside Passage to the south all intersect. Here, the Aleutian Low and North Pacific High intermingle, and as climates converge, northern temperate and subarctic environmental conditions overlap. And here, representatives of all of the major Native groups in Alaska had converged to live and trade: the Chugachmuit Eskimo, Eyak Athabaskans, Aleuts, and Tlingits.

It is a world unto itself. Prince William Sound is delineated not just by the coastline, the way it indents at Cape Puget and at the Copper River Delta into what, upon his visit in 1899, John Burroughs called "the enchanted circle." It is contained by a string of mountain peaks, among them the highest coastal mountains in the world, some of them nunataks, jagged spires that jut through an ice field. These mountains enclose the Sound, hold clouds and rain in. They encircle it like sentries. The mountains, the water, the ice fields—it seemed to me they were all protecting this place, guarding it from harm.

©

After the trip to North Carolina, we returned to Alaska for good. We left behind family and friends, the home we'd made in Chapel Hill, our lovely ten acres of woods, our gardens, our orchard.

We didn't sell the land. Though Andy thought we should, I couldn't do it. My herb garden was nearly gone, the ground returned to the hard orange clay. Only the mint, the bayberry, and the wormwood still grew. In the lower garden, all was gone to weed except the small patch of asparagus. I knew that leaving the place to renters would only make it worse, but I wasn't ready to let go entirely of the life we had begun there. Still, the pull of Alaska, of Prince William Sound, was too strong to resist.

I did bring back one plant: a potted Christmas cactus. I had started it when I was fifteen years old from a piece that had fallen off my mother's plant. Mine was still not as big as hers, which she lugged outside and set under a forsythia bush each summer. But it was big enough to take up all the room on my lap, all the way back. I wondered how it would grow in such a cold, dark land; I wondered if it would still bloom at Christmas.

We moved into Anchorage, as close to the Sound as we could get and still find work. I got a job as a research assistant at the University of Alaska's Institute of Social and Economic Research. Andy found work at a new golf course. We bought a fifteen-foot inflatable like Peter's to get out in the Sound, while we lived in a twelve-foot travel trailer on the edge of the golf course. Our truck camper became my writer's room.

In the middle of winter, in our tiny metal home heated with three electric space heaters, several feet of snow covering the ground outside, we examined charts of the Sound. Could we carry enough fuel to make it to Eshamy Bay? Might there be a tender during the commercial fishing opener that would sell us a few gallons? What about heading up into the long arm of Unakwik to see that tidewater glacier? Or Knight Island, along that meandering shoreline?

Meanwhile, every Sunday when she called from North Carolina, Andy's mother asked us, "When are you coming home?" as if she couldn't hear the sound of our voices when we spoke of this place.

It's been said that the face of a new love is like a mirror in which one sees the divine light of oneself. Perhaps the Sound showed me that self I was most in need of finding. Perhaps my experiences in this new place, the images I held in my heart, reflected back out into the world, throwing light on that which I needed.

The August night in Deep Water Bay, when I saw the stars for the first time in months. The passage across Port Wells in deep fog, when the dark shoreline appeared in a shaft of light. The evening in Blackstone Bay just offshore Beloit Glacier, when three terns flew overhead, their bellies a brilliant gold of reflected sunlight.

Love isn't a given; it isn't even expected. It happens quietly, when you think you are up to something else. Then of a sudden you feel the pull.

Morning arrived at Decision Point with more of the same: rain. Lying in my sleeping bag, I wished fleetingly that Peter would return and tow us back to Whittier. But I roused, crawled out, and faced the shore. The water was calm, dimpled only by raindrops and stained dark gray by clouds. As we cleaned up camp, a fog settled in, obscuring the view in every direction. We slipped our kayaks easily into the water, making slight ripples that radiated out across the water's glassy surface, disappearing into the fog or washing up to the rocky shoreline of the headland. It was low tide, and the intertidal medley of plants and animals lay exposed. The mustard-green popweed was saturated with color and light; the lime-green sea lettuce glowed.

I glanced back at our camp as we paddled away. Our tent sat nearly hidden in a forest. More shades of green than I could have imagined were in that bit of forest and shoreline, from the dark spruce to the pale lichen on their trunks, from the brilliant sea lettuce to the smoky leaves of blueberry bushes. All was quiet, sound dampened by fog, save my paddle dipping into the water. The spruce trees glistened with fog droplets, their tops shrouded in mist.

The rain quickened, pattering louder on my raincoat. Fog swallowed the shoreline, and I paddled hard to keep up with Andy, falling into step behind him as I might follow another car down a road on a moonless night. My eyes focused on the stern of Andy's orange kayak; my mind wandered away from the cold and the wet.

The shoreline revealed itself when we were only a few yards from it. As we beached the kayaks and pulled them up above the incoming tide, the fog pulled back before us, revealing a band of grayish blue: Tebenkof Glacier. Above the glacier the sky cracked, and a blue wedge grew larger and larger, seeming to pour from the ice field down the glacier's icy tongue toward me.

But Tebenkof was no longer a tidewater glacier. Its landlocked face was now protected by a band of dense alders, tenacious pioneer plants. We tried several routes through the thicket, but turned back each time, scratched and frustrated by branches that clawed our arms and concealed our view. We contented ourselves with a beach walk, then headed back across the bay, back into the fog.

Andy, once again in front, started paddling out of sight, and I called to him to wait up. I wanted to fall into the trance of following his stern again; I wanted his company while crossing this wide expanse of water with no land in sight; I wanted him to know that I needed him. His pace was so much faster than mine that I finally gave up. He sped off.

I stopped to watch water drip off my yellow paddle into the dark water—deep water, cold water. The shoreline of these glacial fjords drop sharply to depths of hundreds of feet. Temperatures rarely exceed fifty degrees, even in late summer; hypothermia could set in within minutes of immersion; within five minutes, I'd be too numb to save myself.

A thousand feet of icy darkness lurked just below this thin blue layer of fiberglass. My kayak was no bigger than my body, just as long and narrow, and lighter. It sat so low in the water, I felt as vulnerable as a surfer in shark-filled waters. Just a thin blue piece of fiberglass, like a sheet of paper, between me and this vast ocean world.

Below me, in the deep waters, there could have been salmon. All five Pacific species swim the waters of the Sound. As do three-hundred-pound halibut, black cod, rockfish, Pacific herring. As do whales—minke, humpback, sei, fin, killer. There could have been a thirty-five-ton humpback whale beneath my kayak, one of a hundred that spend summers in the Sound, feasting on krill that drift in currents. Weeks earlier, I had seen a photograph of a massive humpback-whale tail raised behind a single kayak, looming over the small boat ominously. Though I'd heard that whales somehow avoid hitting boats, it was easy to imagine that tail crashing down on this toothpick of a boat.

Or maybe a smaller marine mammal slid beneath my thin kayak: harbor porpoise, Dall's porpoise, Steller sea lion, harbor seal, sea otter. Those two kayakers I had met in Whittier had encountered a group of sea lions at Icy Bay, and a bull tried to climb up on the bow of one kayak. The two-ton animal would have rolled the boat.

Such life swirled below, out of my view. These waters teem with life, regardless of the cold and dark. Fed by huge plankton blooms from the Gulf of Alaska, these are some of the most biologically productive marine waters in all the world—an array that overwhelmed me as I sat in my tiny kayak, floating in a fog. And yet, all I had seen was one solitary jellyfish, a pale-orange cloud of life pulsating by with a languidness that showed no fear of death.

"Pfffhh." A sound, rocket taking off, air expelled. "Pfffhh." Whale.

"Whale!" I shouted to the space in front of me where Andy should have been. "Killer whales!" I shouted as another surfaced a few yards away, enough to see black and white and fin. "Andy, killer whales!" My voice was high-pitched, my breath came in gulps, my heart beat like a caged bird trying to escape.

"Pfffhh." The sound was farther away; they were moving past me. "Pfffhh." I saw another surface, parting the water. It was too small to be a whale. Must have been a porpoise, a Dall's porpoise. Yes. They have the same markings as a killer whale, but are a third the length. And they are fast—among the fastest marine mammals on earth. They sliced the water like scissors cutting paper.

I had seen them once before, playing across the bow of a ferry boat, arcing back and forth. Maybe they saw my kayak and thought I'd speed up and run with them. Still, my heart thumped, and my hands could barely hold the paddle.

Andy finally paddled into sight.

"They're gone," I breathed. "Dall's porpoise, I think, not whales."

"Cool," he said. "How many?"

"I don't know. At least three, maybe ten, I don't know." My breath slowed, I gripped the paddle.

"Are you OK?" he said, surprised at the tremble in my voice.

"Yeah, they just caught me off guard," I said.

But I was thinking about the picture of the whale's tail behind the kayak, I was thinking about the sea lion trying to mount the slim kayak, I was thinking of all I could not see or even imagine, all that was right now around me. And I was thinking of how the unexpected could be both frightening and wonderful. What else lay below that might show itself to me? As suddenly as the porpoises moved from air to water, my fear was sliced through by joy.

I felt Prince William Sound gripping me, and I made no move to wrestle free.

Thunder

Blackstone Bay

We motor as close as we dare, until the jagged face of the glacier fills the sky before us, its shards of icy blue replacing the sky's indistinct blue. Then we cut the motor and sit. Wait. Watch. Our patience is rewarded. There. On the far-north side, just above that granite mound emerging from the ice face, first one chunk plops into the sea, then a waterfall of falling pieces of glacier, a crescendo of sound as more and more join in, and then the rippling of waves out to the boat, the gentle rocking of long, deep waves.

I could go now and be happy for the rest of my life.

But wait. Another, this time in the very center. A column, looking like a frozen bolt of lightning, tilts, breaks free, falls straight down into the gray water, jackknife dive, huge plume of water like a tidal wave pushed up, out, crashing back to the surface, then long, deep waves, those hitting shore breaking ten-foot waves, those toward us rocking, tilting our world, leaving me unsure of which way is up. It is dizzying, the way the motion, sound, motion, come in separate waves, the way each one is different, the way I want to stay here forever.

I have become a glacial erratic. A human left behind to sit in front of a glacier and watch it be. Agape.

PART II

What we see now is death. Death, not of each other but of the source of life—the water. . . . It is too shocking to understand. Never in the millennium of our tradition have we thought it possible for the water to die. But it is true.

WALTER MEGANAK SR., Chief of Port Graham,
Address to the Oiled Mayors of France
and Alaska, June 1989

Do you want to improve the world?
I don't think it can be done.

The world is sacred.
It can't be improved.
If you tamper with it, you'll ruin it.
If you treat it like an object, you'll lose it.

LAO TZU, *Tao te Ching*

struck

1989

Bold black lettering roared the news, a headline in 125-point type, the size reserved for our greatest disasters and greatest triumphs—PRESIDENT ASSASSINATED! WAR ENDS! EARTHQUAKE KILLS THOUSANDS! Lettering so thick I couldn't read it. It blurred before my eyes into a pool of ink. Of oil.

I turned away. In shock or in denial, I didn't know; numbed, I went to work, sought refuge in routine.

But there was no routine. All around me, questions and theories flew like bats startled from sleep. Had the ship's captain, Joe Hazelwood, been drunk? Was the helmsman incompetent? Was it some kind of conspiracy or act of terrorism? How much oil had spilled? How would it affect the beaches? Would it spread to the western side of the Sound? What would they do to clean it up? Should they set it on fire, should they use dispersants, or would that make things worse?

All these questions looming, all these people loitering about my office. My coworkers knew I spent a lot of time in the Sound, knew it was the reason I had stayed in Alaska. So they wanted to know, what did I think of this oil spill? But I had never once imagined that a tanker traveling across Prince William Sound could do this, could spill so much oil, could—in one wrong turn—threaten the very existence of this place. I had nothing to offer that morning but my own despair.

I sat in my chair, silent, astonished by the arguments crackling in the air above me. I was sunk deep in grief, but some of my coworkers were energized, formulating intellectual arguments about this disaster as quickly as the oil itself gushed from the ruptured hull. Reverting to the familiar, they put the oil spill at a safe distance, a distance I simply could not attain. For me, the boundary between personal and professional had washed away; these arguments were meaningless to the death about to unfold.

I gathered the energy to rise from my chair and slip out underneath the cross fire. I headed for the bathroom, seeking solitude and a place to weep. Had I expected the news to not travel to my office? Had I hoped to just slide into work and let the daily routine create a temporary amnesia?

I sat in the stall, holding my head. So it was real. It had happened. An oil tanker had struck Bligh Reef. Ripped the hull. Spilled oil. Was spilling oil. Even now, this moment, oil spreading, animals dying. And this moment, too. Now. And now.

I emerged, tried to work, headed for the bathroom again. And again. My chest ached, threatening to burst from the tidal wave inside, and I couldn't think. My throat burned as if it were on fire, and I couldn't speak. Couldn't comprehend this truth. Couldn't push it away, couldn't take it in. I told my office mate I was sick, I had to go home.

I was home for a week.

Three days. Three days of confusion, of still waters, of the oil pooled around the crippled *Exxon Valdez*. No one seemed to know what to do. Not the U.S. Coast Guard, not Exxon, not the State of Alaska. Alyeska—the consortium of oil companies that owned and operated the Trans-Alaska Pipeline and terminal—had brought their contingency plan for responding to oil spills off the shelf, wiped off the dust, cracked the cover, but still taken no action. Then the wind picked up, and the oil fanned out and spread across Prince William Sound.

Four days. I kneeled on the living room floor and watched maps on TV, hands pressed together. I prayed it wouldn't spread to Knight Island, and it did. To Culross Passage, and it did. To Port Nellie Juan, and it did. It would not stop. It grew greater each day. Thirteen hundred miles of coastline coated in oil. Poisoning, suffocating, dying all around. All the places I loved. All the animals I'd met there. And inside me, too, a poisoning, a suffocating, a dying. I paced the house, restless, lost.

Five days. I was supposed to write a radio commentary about Edward Abbey, who had died of a heart attack on March 14, ten days before the oil spill. But how could I write of one man's death when Prince William Sound was smothering under a spreading cloak of darkness? What could one life possibly mean in the face of this horror?

Six days. I knew what to write. I sat down, and the words flowed from my hand like liquid. Edward Abbey's heart had nearly broken when Glen Canyon was flooded to create Lake Powell. Even though his most precious place was lost, his words continued to give it voice.

Oil now flooded Prince William Sound. It was lost, changed, if not forever, then at least for my lifetime. I feared my heart was breaking.

I wrote a eulogy. For him, for Prince William Sound, for us all.

Goat

Culross Passage

Look. Out across Culross Passage. Halfway between here and the shore of Culross Island a mile away. Something moves. Is it a sea otter? Probably; it's the only marine animal that swims with its head out of water for long. But it swims too clumsily for an otter, a humping dog paddle rather than the smooth, steady grace of an otter. Human? But we haven't seen another boat for two days. Andy and I are here the summer of the oil spill, late in August, when cleanup operations have shut down for the season. There's no one else around.

Only this creature swimming toward us. It moves like a child learning to swim, cumbersome, almost flailing. Moving so slowly, laboriously. It gets closer, and we see a furry head. White. Maybe an old sea otter, maybe it is hurt, or poisoned by oil so it moves awkwardly. There's still so much oil. It took three tries before we found a beach not littered by baseball-sized tar balls and escaped absorbent pads, thick white paper towels stained black. And we've ended up with some oil splotches on our bow after all, in spite of all our precautions.

The swimming animal gets closer and becomes more distinct. White head and white back hump. Thick, wiry fur matted with water. Two small black horns the same size as the ears. Dark eyes barely visible beneath the fur, a narrow face and snout, black nose. It's almost to shore before we identify it: a mountain goat. Must be. But what is it doing here, swimming across the passage?

The mountain goat climbs out on the beach a few yards from us. The goat sees us and stumbles onto a small outcropping of rock. With the fur matted down, her wet, bedraggled body looks surprisingly thin. She shivers and her legs are wobbly, unstable, fragile sticks leading to narrow jet-black hooves. She is so vulnerable.

She stamps her feet furiously, first front, then back. She shakes her head like someone trying to stay awake. Stamp. Shake. Stamp. Shake. Forcing circulation back into her numbed body. Shaking enough water from her fur that she no longer looks as if a good breeze could topple her. The goat stamps and shakes for nearly a half hour. Then the trembling stops; the hooves are steady on the rock. The mountain goat turns and scrambles up the rocks into the forest, disappearing like a phantom.

And though later I'm told by wildlife biologists that we couldn't have seen a mountain goat there, that they don't live on the island and they don't come down to saltwater and they definitely don't swim, even still, I know what I saw.

blood on the hands

1989

I'M TEN OR TWELE, walking in downtown Ashe-ville, North Carolina, with my sister Ginny after shopping. We see a small song-bird huddled in the doorway of an empty shop. We approach it, excited that it is letting us get so close. Then we realize its wing is broken—it can't fly off. Ginny dumps her new shoes out of their box, and together we try to catch the bird; we will put it in the box and take it home, we will heal it, then we will return it to the wild air. We move in; we miss. We're surprised at how hard it is to catch. Three times we try; three times we fail. To escape us, the bird has hopped into the middle of the road. We're scared. We hear a bird cry overhead, see another small bird swooping low toward the wounded bird. It's the same kind of bird; is it a mate? A car rushes by and hits the broken-winged bird, but it's still alive. The mate cries more frantically, swooping lower, as if to try to grab the bird and carry it aloft. Ginny and I are shrieking, too, and crying. Before we can do anything, another car runs over the bird, flattening it. The bird overhead continues its shrieking and swooping. We are crying, helpless now. All I can think all the way home is, if only we had left it alone!

◎

Moose starving in a harsh winter, hundred-year-old trees falling to a hurricane, three whales trapped in the closing lead of an ice sheet: all these natural disasters saddened me, but I had long ago decided to put my faith in a plan that was larger than I could see. I had saved a few birds from my cat's claws, but that was different: my cat had the advantage, a well-fed pet aided by illusory sliding-glass doors.

No plan I could conceive of, though, explained the more than eleven million gallons of crude oil from the *Exxon Valdez* tanker that had spilled into Prince William Sound. I felt as stunned as the birds that used to hit my sliding-glass doors. As the days passed, I became restless and angry. This was not a natural disaster; it was human-caused; it was avoidable. It was different.

Lying awake one night two weeks after the tanker had grounded, I pounded the pillow in anger. I heard on the evening news that the death toll for sea otters was difficult to gauge. The smallest of the marine mammals, sea otters depended upon their fur, not blubber, for insulation and buoyancy. They lost both when oil coated their fur. They died of hypothermia and ingested oil, then sank to the bottom rather than floating and washing onto a beach where they could be found.

Every time I closed my eyes, I was haunted by a single image: a sea otter on the water, rubbing and licking its once silky fur in attempts to rid it of the thick mat of dark, sticky oil, flailing in water it once floated easily upon, shivering and shaking and finally dropping out of sight. Every time I sunk my fist into the pillow, I imagined another sea otter sinking like a stone into the cold, deep waters of Prince William Sound.

Andy was staying up late regularly to watch the news for updates. I didn't know if he watched out of curiosity or because it helped him deal with his feelings. He sat with the newspaper, in front of the TV; I never saw him get upset—upset as I was, constantly. One shot of an oiled cormorant staring bewilderingly into the camera, and I would break down in tears again.

From the beginning, we had reacted differently. That terrible morning, I had awakened to Andy's voice.

"Guess where the biggest oil spill in the country is."

I didn't answer; I didn't want to wake up.

"Guess," he said again.

I don't care, I wanted to shout.

I hated when he did this, when he read the paper aloud to me, jarring me with world news first thing in the morning. I liked to wake up slowly, to savor that naked moment of just awakening, when I did not yet remember what I had done or what I must do. When I simply was, and the world was all possibility.

He knew that. But he stood there.

"OK, where?"

"Prince William Sound."

Anger rent the veil of sleep. Anger: at Andy for waking me with such horrible news. Anger: that this could have happened. That one of those oil tankers traversing the Sound every day could wreck and kill. Anger: that I could have been so naive as to not suspect this possibility. That this place I loved could be so defiled.

Night after night I was tortured with thoughts of oiled sea otters. I had to do something. I couldn't just go on with my life as if nothing was happening.

Still, an old doubt nagged me. The French philosopher Voltaire had said that the purpose of medicine was to entertain the patient while Nature cured the disease. Though this was not a natural disaster, I wondered whether the Sound might heal faster if we just left it alone.

But my restlessness grew. In mid-April, I drove to Valdez on the eastern side of the Sound. The town was already so overrun with people that it was difficult to find a place to stay. Along with a full contingent of government and oil company officials were throngs of media and hundreds of individuals who felt compelled to come—out of a desire to help, but also out of curiosity. The town of four thousand had doubled in size.

My friend Naomi knew of a family who was letting reporters stay at their house, so I told them I was a freelance writer. There were so many reporters that it was easy to blend right in, even if I didn't have an official badge from the *Washington Post* or *Time*.

Driving down from the snowy heights of Thompson Pass into the low-lying valley that held Valdez, I expected the town to look as altered as the oil-covered beaches. At first I noticed nothing different; it was the same pretty coastal town I'd visited several times before.

With clearly laid out roads, distinct residential areas, and a cluster of businesses, Valdez had always appeared neater to me than other coastal Alaska communities. Rather than being crowded between mountains and shore, like Cordova or Juneau, Valdez stood on a flat expanse framed by mountains. It had been rebuilt, moved to a safer site farther inland, after near-total destruction by the 1964 earthquake.

Valdez was more affluent than most small Alaskan communities. Many of the houses were larger, and public buildings like the Civic Center seemed sized for a bigger city. The city's tax base was ample, with the Trans-Alaska Pipeline terminal employing such a large number of residents that Valdez was often referred to as a company town. But it also had an active commercial fishing fleet, and a growing tourism business, both of which stood to lose terribly from the effects of the oil spill.

This oil spill was the commercial fishing community's worse nightmare, one that they had feared ever since the pipeline route was proposed in 1970. At the time, there were two choices: run the pipeline to the Valdez port, and then ship it across Prince William Sound, or build the pipeline to connect with an existing line in Canada, and make it an overland route. The commercial fishing community and environmental community had lobbied hard for the Canadian overland route. A heated argument had ensued in the United States Congress as others argued that the Valdez route was cheaper and quicker, and protected national security and national control. In the end, the deciding vote had been cast by Spiro T. Agnew, who was later discredited and disgraced—but not before putting the Sound at risk. U.S. senator Ted Stevens assured Alaska's commercial fishermen that he had been promised the technology was so advanced that "there would not be one drop of oil spilled into Prince William Sound."

As I drove closer to the center of Valdez, I saw two men with large cameras in their hands and press tags around their necks. The traffic seemed far worse than during the height of summer tourism season. Still, it must have been calmer than it had been those first couple of weeks into the spill, when disagreement among several news photographers exploded into a fistfight over who would get to photograph the first oiled bird.

I drove straight to Naomi's friends' house, but no one was home. I parked near the docks and walked around. Freight pallets holding oil-absorbent towels, piles of blue fish totes, and rows of clean white boom—a synthetic curtain to contain the oil—crowded the narrow wooden walkways. Two fishing boats were entering the harbor; a small group stood talking at the far end of the dock. Their voices were sharp on the cool evening air.

Commercial fishermen in all the spill region's communities, but especially in Valdez and Cordova, were arguing with one another over whether they should sign on with Exxon for cleanup work. It was so highly divisive that old friendships splintered. Those who signed on spoke of "going oiling" in the same way they once said they were "going fishing." Some of their remarks were so inflammatory that I could see vendettas arising and persisting for years to come.

"The money's so good, we should have an oil spill every year," one had said.

A local woman, fed up with the greed of those gone oiling, had commented, "First we had pollution in the Sound; now we've got it in men's heads."

I left the docks and walked to the animal hospital where oiled birds and sea otters were taken. I hoped to be of some use there.

All around the small tan building were freight pallets, walls of blue tarps, buckets, hoses. People wearing plastic bags and rain suits bustled about. Inside, the walls and floors of the hallway were covered with plastic.

I met Cathy, a volunteer veterinarian from Michigan, as she hurried down the hallway. She told me to come back during her midnight-to-eight a.m. shift. Michael, another volunteer, told me they didn't need any more volunteers here, but they did in Seward. The sea otter effort would be moved there. He also suggested I come back at noon the next day for a press tour of the bird center.

Then he showed me the makeshift offices of the Alaska Department of Environmental Conservation (DEC). Security guards in the hallway were posted to keep us out if we didn't have passes. Michael did, so we both went in. There, I signed a waiting list for an extra seat on one of the many flights going to the spill area. I wanted to get out on the Sound and see the oiled beaches for myself. Maybe I could figure out some way to help if I could understand what was going on. Or maybe, like Thomas Aquinas, I needed to see it myself to believe it—oil on the beach as real as blood on the hands.

At two a.m., I returned to the Sea Otter Center. Cathy secured my entrance by telling the guard I was a photographer with the *Minneapolis Tribune.* Inside, it was quiet and dark, too dark for pictures. Cathy sliced crab, clam, oyster meat, using her thigh as a cutting board. The pungent juices soaked through her pants leg. She fed the morsels to two sea otter pups. Sometimes they took the food, sometimes they didn't. One, holding a scrap of blanket close, began squealing. Cathy tried to give him more food, but he wouldn't take it.

"Otters are such gregarious creatures," Cathy whispered, "that sometimes just looking over at them makes them calm down."

She tried this, but still the sea otter squealed. She reached into a cooler, brought out a chunk of ice, and offered it to the pup. Yes. He gnawed on the ice, still clutching the blanket scrap.

I wandered back outside where another volunteer watched several adult sea otters in a plastic swimming pool. He was upset about some of the employees that Exxon's subcontractor, Norcon, was paying to watch otters.

"They just lean on the pens and stare at the otters, talk to them like they're dogs," he said. "The otters were very upset when I came on duty tonight. They have been bothered too much, and they were pissed, making hissing noises."

Back in the maze of rooms and makeshift compartments, I found Judy, another volunteer veterinarian from outside Alaska. She watched over a harbor seal pup that had been so heavily coated with oil that rescuers had thought it was a sea otter. Born three weeks premature, the pup still had its umbilical cord. The tiny seal lay as still as a newborn, eyes squeezed tight, twitching a bit. Another volunteer kept the small webbed feet warm with a towel soaked in warm water. In the morning, the pup would be sent to an Anchorage animal clinic.

All night long, I tried to stay inconspicuous and out of the way, watching while these experts from other places doctored and guarded the wild animals of Prince William Sound. I felt gratitude to them, for their knowledge and dedication and compassion. I was also acutely uneasy, though I didn't know why.

The next morning, I went back to the DEC office. No plane ride today, so I filled out a form to go out the next day. As I left the building, I met a reporter from the *Los Angeles Times* who told me the forms meant nothing.

"You just have to show up at seven a.m., and it's first come, first serve," he said.

The next morning I did just that. A line of ten or so people filled the hallway. I queued up and asked the man in front of me what we were supposed to do.

"Who are you with?" he asked.

"No one, really. I'm a freelance writer."

"Just tell them you're with Ops, then."

I took his advice, and within an hour I was climbing into the front seat of a helicopter. The pilot turned to me and asked, "So, what do you want to see?"

I had no idea. I thought I was just taking up a spare seat on someone else's scheduled flight. Evidently, "Ops" people had priority.

"Take us to some heavily oiled beaches," I said.

First, we took a reporter for *Rolling Stone* to the village of Chenega Bay on Evans Island. Along the east side of Knight Island I had my first glimpse of oil on water—a rainbow of reflecting colors spreading across the water, radiating up into every cove and bay. At Chenega Bay, concentric patterns of boom led in tighter and tighter rings to the Sawmill Bay pink salmon hatchery.

The pilot flew on to Herring Bay, large and heavily oiled. In a desperate attempt to contain the spreading slick, oil was being diverted into Herring Bay, sacrificing its beautifully scalloped beaches. Some areas were boomed off; others were congested with fishing boats and a few barges. All the boats were clustered together in certain areas; if I hadn't known better, I would have thought they were massed for a halibut opener or a salmon run. Flying low over one, we saw two men in rain suits raking lumps of oil into the fish holds of the boat, holds that last summer had held salmon.

In a smaller bay, we saw a raft of six otters and a few pairs of sea ducks. The water around them appeared clear, but I knew they were trapped by the oil. Would they stay in that bay and survive? Or would the oil advance and spoil this small sanctuary? Would they swim out into the slick? Later, back in Valdez, I heard that some people, in efforts to capture oiled animals, were unwittingly chasing clean birds and otters into oil.

Farther down the coastline of Knight Island, the pilot landed the helicopter on a small beach. It was thick with crude oil, slushy and odorous. The sludgy waves made a strange muffled, thumping sound as they hit the beach; the air was filled with an acrid, sharp scent. In minutes, my head hurt and my eyes stung.

As we stood there next to the helicopter, staring at the oil, a man who accompanied us said to me, "I'll take your picture with the oiled beach if you'll take mine." I forced a laugh, thinking surely he was kidding, knowing that jokes about the spill were spreading as fast as the oil itself. Standing there in that surrealistic scene, so shocked by it all that I couldn't trust my senses, I did not raise my camera.

As the helicopter lifted off again to head back to Valdez, turbulence from its wings beat the oiled water into an iridescent rainbow, shimmering wave sprays. At the edge of the sprays bobbed two sea otters, staring up at us with eyes that no doubt stung worse than mine. A small boat cut through the advancing spill; its wake parted the oil, creating for a moment a swath of blue water.

On the fourth day, I left Valdez with nothing to show but a notebook full of words and a head full of shocking images. The chaos exhausted and frightened me. I had never seen so much boat and plane traffic in the Sound, so much human activity on so many beaches. Later, I learned that hundreds of aircraft, more than a thousand boats, and more than ten thousand people were involved in the cleanup. It didn't feel right to me; it didn't feel real.

I wanted to get home to Andy, to tell him about all that was going on in the Sound, in the bays and fjords and beaches that he and I had once enjoyed all to ourselves. I wanted to talk with someone who knew it as I did, who knew the Sound before the oil spill.

On the seven-hour-drive home, I thought of how exhausted and frightened the remaining wildlife must be, of how they could not escape it as I could. They had no refuge. I feared for them, both because of the oil and because of the cleanup. I wondered what we were doing out there, to the animals, to the place. I wondered if we were helping or making it worse.

At the Valdez animal treatment center, I had taken a picture of a sea otter who had been cleaned and awaited transport to a larger tank in another town. He peered up at the camera through thick wires, huddled in a cage no bigger than a shrimp pot. What's not in the picture was what he was probably most aware of: the rumble of a forklift moving crates; the roar of cars and trucks driving by; the waves of chatter from people flowing by to look, photograph, even touch him; the sunlight bearing down on him in a broad dirt lot, no trees or shrubs or water to provide shade, protection, sustenance.

Sea otters weren't land animals; he lived his life on the water, dove beneath its surface to find food and dodge predators. How frightening his tiny, lonely, landlocked cage must have been for him, how far removed from the clear, cold waters and crystal ice floes that were his world. I doubted he survived, but his image was burned into my memory.

൭

Back in Anchorage, I went to a town meeting held at a small Russian Orthodox church. There, dozens of people gathered to find out what they could do. It was one of many meetings hastily thrown together and filled with people desperate to do something, anything, to stanch the flow of death and destruction.

One man, lean and grizzled and looking as if he, too, were having trouble sleeping at night, stood before the small altar, bare except for a white cloth.

"They're not cleaning up the oil fast enough," he said. "They say there's not enough boom. I say, let's get some straw and a boat and go out ourselves. We can cover beaches with straw to absorb the oil."

"I've got a skiff," another man yelled, jumping from his seat. "It's just a twelve-footer, but it can hold two people and some straw."

"How do we know straw works?" a young woman in the back asked.

"Well, I heard they used it on the beaches in that Washington oil spill a few years back," the man at the front replied, squinting into slanted sunlight.

"Look," said a woman standing off to the side, whom I recognized as the director of a local environmental group, "I know you all want to do something. But more people and boats running around out there won't help. Especially not if someone gets in trouble and has to be rescued."

"But we've got to do something!" the man in the front shot back, his voice rising. "They're not doing enough; it's up to us."

"Yes, maybe they're not. But it's better to proceed cautiously here," she said, her voice calm and even. "Why don't we call the folks in Washington and see if straw really does work before we plan a trip."

"Well, I don't know . . . ," he said, and she pulled him aside, talking quietly to him.

I knew she was right: more people out there wouldn't help. I knew he was right, too: we had to do something. At least we could all agree on that.

Meanwhile, others approached the altar and offered their ideas about what we could do. But as the shadows lengthened on the full pews, the talk ran itself out, and the focused energy seemed to dissipate into the old, dark wood. No single idea had carried the crowd, no single idea had shifted the weight that hung over us all. The meeting slowly disintegrated. People milled around, talking angrily, crying, their faces mirroring the thunderhead of grief that gathered in my chest. A room full of people with so much energy and nowhere to put it. I slipped out the side door and went home, more lost than ever.

©

A friend I'd worked with in Chapel Hill sent me a postcard. She knew I was working on a career in writing. She had heard my commentary on the oil spill on National Public Radio, and wrote to congratulate me.

It felt good to hear from her, to know she had heard my commentary. But I was uneasy about a writing success that came from such a disaster. I felt guilty, like an ambulance chaser.

Another friend from Cordova also heard the commentary. She told me she had been in her car, driving to a meeting about oil spill cleanup efforts. She told me she had to pull over, it made her cry so hard.

Thinking of how these two friends had responded so differently, I wondered how the rest of the country, the rest of the world, was reacting. I knew they were being subjected to many of the same disaster images: people in what looked like orange space suits washing black beaches with high-powered hoses; a pile of oiled bird carcasses awaiting sorting, counting, and storage in a freezer truck as evidence in the lawsuit against Exxon; a brown bear on the beach, half his face black.

Hurricane Hugo blasted through North Carolina that fall, one of the greatest natural disasters to ever affect the United States. When I talked with family and friends, they told me about the damage to houses, trees, and streets. I could hear my mother-in-law's grief over the toppling of walnut trees her parents had planted when they built their first and only house. I worried that the large oaks on our land in Chapel Hill might have also fallen. I could tell the damage was massive.

But I began to wonder if they shared their disaster stories as a parallel to what was happening to Prince William Sound. To say that, *Yes, we understand. We have a disaster here, too.* I wanted to tell them that not only was the magnitude of devastation greater, not only were the long-term effects of poisonous oil greater, but while the hurricane was natural, the oil spill was not.

I wanted them to know that it felt different. That the oil spill had been caused by people, that the oil spill could have been prevented. That there was anger and blame mixed with the sorrow and grief. But I didn't know how to tell them. I barely understood it myself.

At the town meeting, I had put my name on a volunteer list. A week later, I was called to go to Homer, 220 miles west of Prince William Sound on the Kenai Peninsula, to capture oiled birds. I was torn about whether to go; my experience in Valdez left me worried that we might just get in the way, even do more harm.

I met up with two others who, like me, had no training with birds but were compelled to try anything. One I didn't know: John worked at the Alaska Public Lands Information Center downtown. The other, Steve, was an economist who worked with me at the university's Institute of Social and Economic Research.

The three of us drove the five hours to Homer and arrived in late afternoon at the volunteer office. Five other people crowded the tiny office, none of whom knew we were coming or knew what to do with us. We waited on the stairwell, trying to find out how we could get out in Kachemak Bay and rescue oiled animals. That was what we had been sent to do, but except for a couple of nets propped against the office wall, such a plan didn't look possible. We had no boat. After several hours, we headed to our motel room, frustrated, tired, and confused.

The next morning, we met Mei Mei, a local writer who had entered the chaos and emerged as coordinator for volunteer efforts. She was quick and efficient, her short dark hair and dark eyes intensifying the effect of someone working at top speed. We hurried to follow her across the parking lot, trying to catch the few spare words she tossed over her shoulder. She showed us the high school swimming pool where a couple of oiled otters were being held, the Quonset hut where oiled birds were being kept warm and fed, and the office building where dozens of people rushed around, talking urgently, making copies of large documents, answering dozens of ringing phones.

"This is the nerve center," she said inside the office, just before she was pulled aside by a tall, thin man. They began talking rapid-fire about the logistics of getting the birds to a treatment center in Valdez versus setting up a treatment center in Homer. Most Homer folks wanted to set up something here, keeping the animals close to their home and saving them the added trauma of plane transport.

We stood around, waiting for Mei Mei to finish her conversation. Steve finally said, "Let's just head back over to the volunteer office and try to find a boat to take us out."

Since we couldn't discern much but administrative confusion and chaos around us, John and I agreed. We thanked Mei Mei, she gave us a quick nod, and we returned to the tiny office.

"We've found you a boat and captain. You can go out in the morning," a woman who appeared to be in charge told us. "Here, take these nets, and go to the building next door at four p.m. for training."

We grabbed the nets on the stairs and left. After a rushed late lunch, we found the classroom where the class was to be held.

I found a spot on the floor of the meeting room where twenty or so others already sat, waiting for our instructor, Charlotte, to arrive and teach us about birds: how to catch them, how to hold them, how to feed them. How to save them. We sat quietly on a thin blue carpet, in a room with blank walls and no furniture. I kept raising my head and staring out the only window, which let in light and green trees, branches and a bit of blue sky. Like a murre or cormorant might, caught and crouching in a cardboard box, peering up through the crack of light in the lid, waiting for freedom, willing it with an unwavering stare.

Early the next morning, we were quickly paired with a boat and captain, and after a scramble to find enough life jackets for us all, we were off.

It was a small fishing boat, crowded with all five of us aboard, and a bumpy three-hour ride westward out Kachemak Bay in big waves. We headed to Flat Island at the mouth of Kachemak. We had heard that hundreds of oiled murres were being blown there from their nesting colonies on the Chiswell and Barren Islands around the cape in the Gulf of Alaska.

Passing Jackalof Bay and Seldovia, we began to see them. We watched for signs that they were oiled. A few we circled closely, but when they flew off we knew they were OK. The oiled ones, we had heard, couldn't fly. Feathers matted, they couldn't fly or stay warm enough, and so they most often died of the cold rather than starvation or poisoning.

We began to see more and more murres, small rounded birds, black with white bellies. They look more like penguins than any other bird in Alaska. When penguins appear in drawings of the Arctic, I wondered whether the illustrators didn't know that penguins don't live here, or if they had just drawn murres badly.

Murres stay out to sea most of the year, coming ashore in summer to nest and raise young. If this oil spill had occurred in the fall or winter, the murres, like most seabirds, would have been spared. They form great colonies, tens of thousands of birds, and nest on sheer rock cliffs. Their blue-and-black-mottled eggs are sharply pointed at one end, so that instead of rolling across the ground like a chicken egg, murre eggs spin in tight circles, the pointed end at the center. In this way, a murre egg perched on a high cliff won't roll off as easily if nudged. The birds themselves sit straighter than other birds, appearing to stand up like penguins. They fly fast and low over water, diving like sleek rockets as much as twenty feet into the water to catch the fish they eat. Like penguins and puffins, their aquatic abilities almost outmatch their aerial ones.

These murres we now saw, though, did not flip and dive under water nor flap and fly off at our approach. They bobbed on the surface, thrashing their wings and awkwardly pushing across the surface when we got close. These were oiled birds. They let us get so close we saw the dark stain on their white breasts. We saw how they must have swum right into a slick of oil, and now could neither dive nor fly. Some tried in vain to clean the oil, preening and flipping water over their backs. But by preening they just ingested the oil, a poison that would slowly kill from the inside out.

In the entire spill region, I would eventually learn, three-fourths of all the bird carcasses recovered were murres. Of the colonies that these birds around us were from, nearly half were lost. Ignorant of these grim facts for the moment, Steve, John, and I got out the nets. At first we tried to catch birds from the boat, hoping to scoop them up as we motored close. But they flapped off with astonishing quickness at the last second, and we ended up making large noisy circles in the water as we chased one murre, then another.

I felt a lead weight growing in my stomach.

"Let's try to get ashore, maybe catch them on foot, so the boat won't scare them off," said Steve.

"We could use the boat to scare them toward you," suggested the captain.

Steve and I were dropped off on Flat Island, in our knee-high rubber boots with nets. The boat would corral a bird, herding it toward us. Then Steve and I would stomp around in the shallow water, lurching forward when we got near one. Finally, we found ourselves both pursuing a single murre. The boat had chased her for ten minutes, and she sat, tired and oiled, on a rock just offshore. Steve angled in on one side, I on the other. The boat stayed nearby. Steve lunged and missed. I lunged next, water sloshing over my boot tops, arms shaking and heart beating fast. But I caught her.

She struggled in my hands, her neck lengthening and beak reaching for the skin of my wrist. As weak as she was, as dying of oil, she still did not want my help. With her feet clawing my shirtsleeve, she pulled one wing loose.

"Put your hat over it," I heard Steve yell. So I took off my wool cap and put it over the sharp beak and writhing neck. She stilled.

In my panic, I'd forgotten what Charlotte had told us in bird class. "Cover their eyes and they'll calm down. It's as if, if they can't see you, they think you can't see them."

We climbed aboard the boat, put the murre in a cardboard box, and closed the lid.

"I'm nearly out of fuel. We need to head back," the skipper said.

With one oiled and exhausted bird to show for a boatload of people and a tank of gas, we returned to Homer. Several times on the way back I looked in on her. She made no sound or movement in her box, and it wouldn't have surprised me, after that arduous hunt, to find her dead. I was glad we were returning. I wanted to get her warm and fed and carefully watched by someone who knew what they were doing.

On the drive back to Anchorage the next morning, staring out the window at the forests and mountainsides and still-water ponds, I wondered how the murre was doing. Was she even alive? Would she have been better off dying in peace in the waters of her home? I didn't feel proud about what we had done. And I did not want to catch another bird in that way ever again.

©

The economists and sociologists I worked with held a meeting to discuss research proposals relating to the oil spill. They were debating how to design a project that would assign value to the damage—to the oiled beaches and dead wildlife. One economist described two methods for this "contingent valuation." A sociologist suggested two approaches to survey public opinion on the value of wildlife.

"You can ask the question two ways. One, ask a person how much they would be willing to pay to have the damage not happen. Two, ask that person how much they would be willing to accept to let that damage happen. So much spent to prevent a spill, so much accepted to let a spill happen. You'll get two very different answers," he said.

I sat in the back of the room, listening. I was there only to observe. As a research assistant, and one not specifically trained in these fields of study, I wasn't asked for input into grant-proposal ideas. I had not planned on attending this meeting. My office mate, Eric, had urged me to come, but I wasn't sure if it was for what I might learn, or what they might learn from me. I agreed to go because this was about the oil spill; this was about Prince William Sound.

I couldn't believe what I was hearing. I couldn't believe that all these intelligent, well-educated men actually believed they could assign a monetary value to that oiled murre in Kachemak Bay, to the orphaned harbor seal pup in Valdez. The discussion was completely removed from what I had experienced, disconnected from what I knew of the actual disaster taking place. But who was I to know better?

Eric glanced over at me and gave me a thin smile. He had answered a few of their questions about economic modeling, but had otherwise kept quiet. Was he also feeling uneasy with the direction of the meeting, or was I just imagining his discomfort so that I would not feel so out of place, so alone?

"So, a sea otter might be worth more if a person is getting paid to let it die, and less if a person is paying to keep it from dying," said another economist.

"But you can't," I blurted. "There's no way that some monetary value is going to be enough for one wild sea otter's life. The sea otter is diminished, devalued, the moment you apply a number to it. Unless that number is infinity."

The economist turned toward me. He spoke in an even voice. "Well, a value will be assigned; it's just a matter of deciding how to assign it."

I didn't speak again. The meeting reminded me of the exploitative media in Valdez, of the intellectual arguments in this office the first day of the spill. Could this disaster be reduced to a few research projects, to a few column inches? Was this the best we could do? Such a dislocation between my experience and the responses of others was new to me; it was frightening. But at least now I knew what contingent valuation meant: putting a price on the priceless.

It wasn't more than a few weeks after the Homer trip when I heard that a sea otter treatment center was opening in Seward. The oil was still spreading, having poured out of Prince William Sound and down the Kenai Peninsula, into the fingers of Kenai Fjords National Park, around the bird cliffs of the Chiswell Islands, into Kachemak Bay and down Cook Inlet, past Kodiak, and heading now toward the Aleutian chain. The spill area was now the length of the entire eastern seaboard of the United States. Animals were still dying, and the nightly news was still dominated by images of oil-soaked seals and cormorants and loons, grizzly bears and bald eagles and deer.

I set out again to help fight the tide of oil spill death. The sea otter center sounded organized and efficient, not chaotic like Valdez or Homer. I told myself that would make a difference. I told myself that if Ginny and I had had a better box that day on the street, we could have saved that bird.

Naomi and I drove to Seward together, 126 miles south of Anchorage. The center was built to hold thirty-five sea otters, but had eighty otters three days after it opened. They hungered for volunteers; we were put to work immediately.

The afternoon we arrived, I was given a clipboard and told to watch otters that had been cleaned and swam in makeshift swimming pools. I fed them every few hours and made notes on their behavior every few minutes. In the hour-long training session we received upon arrival, the veterinarians had told us that sea otters are highly social: they live in rafts of up to a hundred, and are hardly ever alone. Sometimes they sleep with their heads on one another so they won't float apart. Here things were different.

"Don't make eye contact," the veterinarians had warned us. "It will make these otters too comfortable with humans, and we want them to stay wild."

This was impossible. The otters persistently tried to get my attention. They swam to the front of the cages; they hopped up in front of me as I walked by; they followed me with their childlike eyes. I could not escape their gaze.

"What should I do? They keep catching my eye," I asked a vet who was walking by.

"Just try to keep your eyes lowered. Try to glance at them out of the corner of your eye. They just want to understand what is happening to them, who we are, so they can feel some security," she said.

Some, however, wanted only to be left alone. In one pen, a young female shivered and would not eat, so the vet had me aim a big blow-dryer at her. The blast of warm air, she said, would dry her matted fur, which had not yet recovered its insulating ability. It worked; she stopped shivering.

But the otter who shared her pool, an older white-headed male, did not like the noise: he quit eating, he laid on his deck, he hissed and looked at me sideways. After a while, we moved the female to an indoor pen, hoping to calm him. When I went off duty at midnight, he was still agitated and turned away whenever I looked toward him.

The next morning, Naomi and I were assigned to wash crude oil from otters. They were all so heavily drugged that they didn't move, except occasionally, when a violent shudder passed through their bodies.

The sea otter assigned to us, a female with a white head, did more than shudder. As I held her head, she pushed up and resisted me. She kicked one hind leg, spraying oily soap on Naomi. We grabbed two other volunteers. Naomi held her feet down, and, with four pairs of hands now on her belly, we managed to finish spraying down her three-foot-long body and working in the soap.

Still she struggled, so a vet came by and gave her another shot of sedatives. Within minutes she was still. The vet then pulled out a hole punch, quickly punctured her right hind flipper, and slipped on an orange tag with the number 73. Naomi gasped.

"It doesn't seem to bother them," he said to Naomi. "There aren't any nerves in these webs between their toes."

We continued washing, checking for oil after each rinsing by stroking her fur with our palms and looking to see if the water was still brown. Her fur beneath my palm was flat and silky smooth, so smooth that it revealed the lines of hard bone and muscle. Her wide body was shaped like a boat, ribs curving up from the backbone to the sides, ending in the soft, slightly concave belly. Perfect for floating on the ocean's surface; perfect for using to eat urchins, crabs, mollusks, and fish; perfect for cradling a newborn pup.

Beside me was another group cleaning another otter. They talked and laughed loudly as if they were at a cocktail party, barely looking at the sedated otter under their hands as they rubbed the fur mechanically. One man told a joke; another woman moaned that her nails were tearing from all the washing. It was considered the latest trend, my coworker Steve told me later, to spend a few days in Seward cleaning oiled otters.

I watched as one man squeezed water out of the otter's tail as if he were wringing water from a wet towel. Was that too hard? The animal was so drugged that, even if it did hurt, we wouldn't know. Would that sea otter wake with a damaged tail, a useless rudder when he would try to swim? Would he be too weak to return to the wild, sent instead to an aquarium in some landlocked city?

We finished cleaning number 73, finally pressing only clear water from her fur, and moved her to the drying room. The handheld blowers we used on her fur were so loud I wore earplugs. Outside, a half-dozen cleaned sea otters in a plastic swimming pool were fed whenever the blowers started, so they'd come to like rather than be disturbed by the noise.

Suddenly, number 73 awoke and began to struggle. One woman laid a towel across her belly for her to bite on. Another shut off the blowers, and we all four stood still and waited. She went back under, and we started drying again. Her fur began to dry quickly, a thick loft of golden brown. As I held her head between my hands, I stood mesmerized by the thousands of fine hairs, the densest fur in all the world, a silken star burst under the blower. The next day, I was still wiping fine otter hair from my eyes.

Once dry, the old female was put in the recuperating room, a quiet, dark place filled with exhausted otters. After the sedatives wore off, these otters were moved to the outdoor swimming pools. If they recovered well enough, they were then moved to a pen in Jackalof Bay, where they would be held until the U.S. Fish and Wildlife Service authorized their release. Most would not survive after their release. Many of those released would first have radio transmitters surgically implanted in their bellies, so that scientists could track their movements and their survival rate. Within two years of their release, all the radio-implanted otters would be dead. Some scientists suspected a herpes-like virus was contracted in the treatment centers, and released otters may have transmitted this virus to other uninjured sea otters.

The debate concerning efforts to save sea otters continued for years. An estimated eighty-five thousand dollars was spent per sea otter. Some said it was too much to spend; others said money didn't matter. Some said that even if only a few otters survive, it was a success, because volunteers learned so much about wild animals and gained empathy through that direct connection.

But all along I thought the reason for rescue efforts was to help sea otters, not to help ourselves. That we were trying to make amends for our mistake, not make more mistakes. When I thought of the trauma they endured in the captures and the cleanings, when I thought of those not well enough to be released being shipped off to aquariums, when I thought of those who were released but died soon after—I wondered if it was worth it to the sea otters.

Every hour or so until I left Seward, whenever the vet would let me, I looked in on number 73. Long after the tranquilizer should have worn off, she was sluggish and quiet. The last time I saw her, she still was not accepting food.

Every day, throughout that first summer, I was afraid that I was not doing enough, that I was not doing the right thing. I heard a call for help, and I went to help. But every time, I left feeling as if I'd failed to answer that cry or, worse, made the cry louder.

When I looked at the faces of the other volunteers, when we talked in low voices around drugged sea otters and dying birds, I wondered if they felt doubt, too. I wondered if they were figuring out, as I was, that this wasn't working. That not with all the technology at our disposal, all the money from Exxon and governments and foundations, all the volunteers pouring in every day could we even begin to fix what we had broken. I wondered if they, too, laid awake at night feeling as I did, that our rescue efforts were more about entertaining ourselves, keeping ourselves busy, making ourselves feel good. That we were living a lie.

Oh, how I wanted, how we all wanted, to fix what we had broken, to make it all right. We wanted to save sea otters and murres and all the animals we possibly could from an oily death. We wanted to clean up their homes, sweep it all away, and then back away with murmured apologies. We wanted to undo the oil spill.

૯

In the back room of the greenhouse, I potted plants. A big, shaggy sod mat next to me held small rosettes of foxglove. I didn't know where these wild plants came from, but we were potting and selling them for three dollars each.

Gray-green downy leaves, shaped into perfect pointed ovals, radiated out from the center like a snowflake. I picked up one corner of the sod mat and untangled the roots of another plant from the mass of foxglove and grass and fireweed roots. Sometimes a root broke, and it recoiled after the snap, as if from pain. But I tried, every time, to keep them unbroken, knowing that each root strand improved the chance that this plant would survive this rude transplant from a shady meadow to this plastic pot.

I was working at DeArmoun Greenery a few hours a week just to be around plants, to have my hands in damp soil, to smell the rich scents of earth—and to learn about growing things in Alaska, for I had no idea what could grow this far north. Now I knew foxgloves did, just as they had in my herb garden in North Carolina.

It felt good to be working this way, in this place. But I had decided to quit. I had worked here only five weeks, and I'd only just begun to learn about gardening in Alaska, but my mind was too often elsewhere, leaving my hands impatient to be there, too.

It had been nearly four months since the oil spill began, and already I had taken three trips to try to help. Still, I was unable to concentrate on work or home or friends—even these lovely foxgloves—for more than a few minutes before I once again found myself out there, on an oil-drenched beach, picking up the nearly unrecognizable carcass of a sea otter or murre or cormorant.

The sod beside me was disappearing, leaving behind a scattering of black soil and a few pebbles. I'd potted ten plants. I stopped to water them, to help soil settle around their roots and to send nourishment up to leaves shocked from transplant. Wherever this sod mat came from, it must have left behind a jagged-edged piece of ground laid bare, revealing dark soil, rocks, roots, untethered and exposed.

"How's it going?" Mary, the assistant manager, had startled me.

"Fine. I've got about half potted, I think," I replied. I wondered if she thought I was potting them too slowly. Such attention to each root takes time.

"Good. They're so pretty, aren't they, even without flowers. Here, take these and write *Digitalis maculata* on them. I think that's the variant we've got here," she said, handing me a bundle of thin plastic strips.

Digitalis. I had forgotten that was the foxglove's scientific name. Why did it sound familiar?

A root snapped. A quick pop. I pulled too hard; I wasn't paying attention. Here I had a living thing in front of me, dependent on my care for survival, and instead I was thinking of a sea otter who was now beyond my help, who may in fact have never benefited from my help.

This was why I had to quit this job and head for Perry Island. I didn't want to capture oiled birds or clean oiled animals again. This time would be different. I would be there two weeks, not trying to do anything except observe. Not trying to fix anything.

Digitalis at my elbow. The mat was gone, reduced to a pile of fine soil, swept up and ready to be tossed into the compost pile. I potted the last plant, minus one root.

Digitalis. I remembered why the name was familiar. *Digitalis* was used to cure heart disease, to cure the heart. At the wrong dose, it could kill. Wild animals left it alone.

◎

In early July, I camped at Day Care Cove on Perry Island with two other observers. Three other groups were camped on other parts of the island: West Twin Bay, East Twin Bay, and Meares Point. Kelley, a Cordova bookstore owner who had spent years kayaking in the Sound, had secured a bit of grant money to provide supplies for this Oil Spill Observer Project.

The weather was unusually hot and sunny, even for midsummer. The cove's rocky beach was still coated in oil, though it had been cleaned with high-pressure hot-water hoses. Within hours of my arrival, I had a headache. The entire time I was there it did not cease.

Our job, every day, was simply to watch and take notes. We walked the beaches; we took the small skiff, nicknamed *Sea Witch* because it was difficult to maneuver, along the eastern shoreline and out to a group of sea stacks and islets called the Dutch Group. We wrote on our small tablets everything we saw.

A solitary sea lion porpoising by Observation Point at 1 p.m.

A fishing boat collecting oiled kelp at the entrance to Day Care Cove at 5 p.m.

A small plane circling overhead at 6 p.m.

No sea lions at the haulout on the northeast corner at 4 p.m.

A helicopter landing at 8 a.m., two Fish and Wildlife agents come to check a bald eagle nest.

An abandoned eagle nest.

A pair of oystercatchers near Aleut Beach at 11 a.m.

An oiled cleanup glove on North Beach at low tide.

A female grebe with three chicks in Day Care Cove every morning.

It was hard, at times, to not stoop and mop up the pools of oil with my bandanna, to not follow the oily river-otter tracks from shoreline into the forest in search of a sick animal.

It was hard, too, to stay on the oil-stained beaches. To deal with oil all around, to keep it from staining my clothes, my tent. Hard to not move to a cleaner beach, to one not so obviously oiled. Hard to avoid denial. And yet essential.

All I did on Perry Island was to bear witness to what was going on in the Sound, recording it for a collective memory, a warning for the future. All I did was record the details of an oil spill, of our feeble attempts to clean it up, of the Sound itself as its remaining inhabitants lived their lives. All I did was to not avert my eyes to the full truth.

All I did was all that I could do.

Seas

Port Wells

We're cutting short our rainy camping trip to Perry Island. The water is rough with six-foot waves, and we shouldn't be out in our little boat. But we can't stand to be wet another day. Everything is wet—tent, sleeping bags, clothes. We stop once, and Andy says we really should wait until the seas calm down before continuing. We remember all the stories of people too anxious to get home who lost their lives trying. But I'm cold and tired, and my period started the night before, made more unwelcome because we're trying to get me pregnant. All I can think of is my nice warm home. We head back out in it, the waves alternately jarring my spine as the bow hits a trough and soaking me even further, saltwater on rainwater, when the bow hits a crest. My eyes are slits, straining to see the entrance to Passage Canal through the wind and rain. Then off to the starboard, out of the gray squall, a sea lion, brown body glistening, pops up, tosses a silvery salmon into the air, pink mouth open to catch it, turns to look at me with dark pools of eyes, then flips under the roiling surface again. I'm astonished at how perfectly comfortable this sea lion is, how perfectly at home.

awakening

1991

DURING HIS FIRST DAYS of life, I barely uncovered my new son for fear he would get too cold. I kept Jamie swaddled in a blanket, his thin arms and legs held tight against the translucent skin of his torso, the blanket folded just as they had taught me at the hospital. It took both Andy and me to give him his first bath—one to hold his wobbling head and unfold his arms and legs, the other to dip the washcloth into lukewarm water and gently pat his skin.

How could I then, six weeks after his birth, take him out into a wilderness known for its cold, wet weather and fatally freezing waters?

When I was pregnant, a friend who had raised three daughters assured me babies can adjust to changes more easily than we think. I was the one, she said, who might not be ready for a camping trip. But when Jamie was born in mid-July, I was filled with energy. My sister Carolyn, who had two sons and a daughter on the way, laughed when I told her.

"You're running on *son* energy," she said.

And so we kept our plans for a late August trip to Prince William Sound. I did what I could to accommodate my baby. Rather than our inflatable boat—small, open, close to the water—we would take a charter. And rather than our two-person tent to keep the rain out and the warmth in, I rented a U.S. Forest Service cabin with a woodstove. This would be our first time using a cabin, our first time using a charter.

Still, it was hard for me to imagine being in the Sound with an infant. I wondered if I was pushing the limits of a baby's adaptability, or of parenthood itself. I suspected that I was; I felt selfish and naive. But I could not keep from going.

We had gone to Prince William Sound as often as possible every summer since we first came to Alaska. These trips weren't just vacations; they were the times I felt most alive, most connected to the world. It was more than the act of leaving behind life's daily details that brought this sense of freedom. It was the place itself, the sound of waves upon a rocky headland, the scent of the forest after rain—the Sound had become my touchstone.

In the wake of the oil spill, my desire to go to the Sound was ignited by a sense of urgency. The place was even more precious now that I had almost lost it. Though I believed it was healing, I viewed everything differently. A cormorant skeleton above tideline, gull feathers crusted with brown along the shore, a lone sea lion at a haulout: were these natural or signs of the oil spill?

I had no idea what the future held—how or if the water, the shores, the animals would recuperate, whether there would be another oil spill, whether something else might destroy them. The forests around Two Moon Bay in the eastern Sound had fallen to clear-cut logging the year before the spill; now the forests along Patton Bay on the outside coast of Montague Island were being destroyed as well. Parts of the Sound I had yet to even see, places I had anticipated exploring, were already being scarred in ways that would take centuries to restore.

I was happy to be a mother. I had made a conscious choice to have a child, and I loved Jamie easily and fully. But I didn't want motherhood to deny me something—some place—I held so dear.

On trips to the Sound, Andy and I had always gone where we wanted when we wanted. We rarely spent more than two days in any one place, choosing instead to explore coves, islands, and passageways. One summer, we headed toward Port Nellie Juan on Prince William Sound's western side, thirty miles by water from Whittier. Wind and rain slowed us down; water sprayed our faces and dripped down our necks; waves made the boat bounce up and slam down with such force that seawater drenched our gear. Still, we reached South Culross

Passage just north of Port Nellie Juan that afternoon. After several landings, we found a beach suitable for camping and pitched our tent.

For the next four days our sojourns were limited only by the amount of fuel our boat could carry. One hot, sunny day was spent climbing a mountain on Culross Island. Another day we explored the coastlines of Eshamy Bay and Crafton Island, returning in the evening under a fog so thick we nearly missed the entrance to Culross Passage.

This time would be different. Two weeks before the trip, when Jamie was a month old, I tried to imagine being in a small boat on the Sound with him. All that came to me, again and again, was a recurring image of him slipping out of my arms and falling into the water, sinking out of my sight into those unforgivingly cold depths.

I was haunted by our friend Peter's experience a few years earlier. Peter had won land in Alaska's homestead lottery, five acres on the shores of Johnstone Bay on the outer Gulf coast between Prince William Sound and Resurrection Bay. He had shown us pictures of dense forest bordering Mount Fairfield and Excelsior Glacier—a dramatic spot to call one's own.

But he could reach it only by boat, so he was carrying cabin-building supplies in his inflatable out of Seward and around Cape Resurrection. On one return trip, Peter came upon two cold, wet boys shivering on the beach; in the water just offshore was their swamped boat and their drowned father. The boat he found was like his own; it was like ours.

The morning we arrived in Whittier to meet our charter was sunny with a wind that might have kept us from going in our boat. I was relieved to have the charter.

We stopped at Swiftwater Seafood, a new little restaurant perched beside the harbor. It was owned and run by our train conductor friend, Dan. He no longer worked for the Alaska Railroad. Dan had throat cancer, and his once booming voice was now a hoarse whisper. Because he had lost his voice, he'd been fired. I was angry at the railroad for taking away his job, too, but Dan didn't seem to harbor any resentment.

He greeted us with the same wide smile as ever, the same open arms and high-swinging handshake. He liked being in Whittier all the time, he said, and was enjoying cooking. He gave us some fish and chips on the house, and waited eagerly for our approval. No question: the conductor could cook.

We walked down the loading ramp to find our charter. The sight of it made my heart sink. The *Sound Runner* was a wide, squat aluminum skiff that was only two feet longer than our boat. It did have a covered front and more space for gear inside, but it fell far short of the tall, comfortable charter boat I had imagined.

What's more, four other passengers would join us. Seven adults plus my baby plus our gear—which included our deflated boat, an outboard, tanks of fuel, and gallons of fresh water—seemed like a dangerously heavy load for such a small boat.

I asked the boat's captain, Gerry Sanger, if the boat should carry so much weight.

"Let's wait and see if she gets up on step," he said, tossing our gear into the bow. "If not, we'll turn back."

Gerry had more years boating in the Sound than any of the handful of charters in Whittier. His was one of only two "water taxis" that would drop off and pick up kayakers and campers. He had worked as a government biologist in the area for many years, and knew a lot about the Sound's wildlife. But now I wondered whether that was good enough for my baby.

Images of Jamie facedown in the water began once more to crowd my mind, like an outtake from a movie playing over and over. Perhaps I should stay behind. I asked Andy; he said it was up to me, but it would probably be fine. Maybe. Or maybe the risk was too great after all. My son was so small and vulnerable, and the Sound was so vast and unforgiving. Once we were out there, help would be a long way off.

Still, my longing to be out in the Sound was so strong it pushed those fears back. I hoped that Andy and Gerry were right. We all squeezed in, and the *Sound Runner* headed down Passage Canal. After several tense minutes, the boat leveled and planed out; as promised, it was up on step. I settled back against my seat.

I was perched on the very end of a bench in the bow, just under the open-ended cover, with Jamie bundled up and held tight against my chest in a papooselike carrier. I strained my back against the jarring ride to keep him from being jostled and to shield him from the wind. Several times water splashed over the side onto my leg, soaking my back and legs, but as long as it missed Jamie I said nothing. Though we passed through some of my favorite landscapes on a day clear enough to see it all, I barely noticed. Instead I stared at the waves hitting the boat and listened to the steady drone of the engine, willing it to run smoothly.

The other passengers, four women from Maryland and Boston, were on a day trip to see glaciers and sea otters along College Fjord.

"How old is your baby?" one asked.

"Six weeks."

"You're brave." She smiled.

"Perhaps too brave," I murmured.

Finally the *Sound Runner* veered toward shore, and Harrison Lagoon came into view. As we motored closer, the cabin appeared among thick underbrush and spruce. The tide was low, baring a wide rock beach punctuated by sun-bleached stumps. Gerry maneuvered the boat through the narrow lagoon entrance, around a crescent-shaped gravel spit, and up onto a rocky beach on the back side of the cabin.

I stepped off the boat; I felt land beneath my feet; I felt Jamie sleeping against my chest, a comforting warmth. We had made it.

The sun shone in a cloudless sky, and calm waters had replaced the whitecaps. I wanted to wander down the beach, climb up into the woods, explore this small peninsula. I also wanted to tend to Jamie, to make sure we had everything set up for his every need. So we spent the rest of the day making the cabin our temporary home. Half our gear was for the baby—dozens of diapers, extra clothes, blankets, hats. I brought too much for him, I knew, but I had no idea which pieces were unnecessary.

Andy and I took turns keeping Jamie entertained and gathering firewood, unpacking sleeping bags and clothes, pulling out cookware and food for dinner. The bare walls and floors of the cabin didn't muffle sound like the carpeted, furnished rooms of our house, and several times noises startled Jamie into crying.

That evening we ventured as far as the gravel spit that swung out into the lagoon right beyond the entrance. In slanted sunlight, we sat on the beach and watched a small flock of glaucous-winged gulls and two young bald eagles circle and swoop over the mouth of a stream on the far shore. A pair of grebes dove for fish in the middle of the lagoon. One eagle, an adult with a black body and pure white head, perched atop a dead spruce on a gravel beach directly across from us.

Seeing these animals was like seeing old friends again. I knew the oil hadn't reached up this far into Port Wells; that's one reason we had chosen this cabin. Still, I was relieved to see these birds, to see apparently healthy bald eagles that had been born since the spill.

That night, the little woodstove couldn't keep the night air out, and I couldn't sleep for worrying that Jamie would be cold. I brought him into my sleeping bag. Tucked under my arm beneath the covers, like a chick beneath his mother's wing, he was so warm he sweated, but he slept well. I slept fitfully, waking often to change positions without disturbing him, or to make sure he was getting enough air.

We awoke to another clear, sunny day. Andy put our inflatable boat together, and I walked with Jamie down to the beach on the Port Wells side. I breathed in the rich sea air; the scene before me seemed to say, *Yes, you were right to come back to the Sound, newborn baby and all.*

Across the water, mountains topped with snow soaked up the morning light. Several sea otters were diving, floating, and eating in the water in front of us. Some were only a few feet offshore, closer than I'd ever been to one. I wanted to sit and observe them, but Jamie cried as soon as we reached the water's edge. The otters dove for safety, resurfacing much farther out.

The boat ready, we headed north for Barry Arm and its tidewater glaciers. On one of our first trips to Prince William Sound, we had spent hours in the boat, basking in the sun and beholding ice falling from the glacier's towering face. Ever since, this was one of our favorite things to do. If Andy and I ever got a bigger boat, I wanted to name it *Glacial Erratic,* because that's how I felt in the presence of a tidewater glacier: like the boulder I'd seen on an expanse of tundra in Denali National Park, a giant anomaly in the geologic landscape that had been deposited far from its source by the river of ice that had once filled the valley. In front of a tidewater glacier, before its majestic profile, I was like that boulder: a rock of a different geologic makeup than all that lay around me, yes, but so solidly planted in place that nothing—short of another ice age—could ever move me.

The closer we got to Barry Arm, the stiffer the wind blew. I faced backward in the boat so the wind would not be directly on Jamie, even though he was completely covered. I zipped up my coat around him and felt like I was carrying him in my belly once more. I could not see the approach to the glacier, could not see the icy face become larger and taller, filling up more of the sky. All I could see was our wake through the increasingly dense field of ice floating in dark-blue waters.

It didn't matter; Jamie was warm and safe.

Andy cut the engine, and I turned to face the glacier. For a few minutes we sat quietly, listening to the ice and surveying its face, waiting for some to break free. The sun warmed my wind-chilled body; I began to relax.

Jamie awoke and began to fuss, whimpering and wiggling in his confining space. He was hungry. I tried to nurse him, but I could do it only by removing my life vest. Around us, icebergs shifted and cracked. One rolled over near our boat, expanding like a surfacing whale as the wake of a calving berg passed under it. Instead of being fascinated by the size of the iceberg, by the light passing through its clear-blue caverns, I began to fear one closer to us would roll and flip our boat. I had heard of a kayaker who was tossed into the water by a rolling iceberg; I again imagined Jamie, wrapped in his brown baby bag, falling into the water, sinking into a tangle of water and ice, my precious bundle gone.

So much below me under the water. So much all around. How was I to protect this child? I gazed into the clear waters and saw an undulating sheen. Before, I would have simply smiled at the water's illusory beauty. Now I feared for my son even as I wondered if it was a sheen of oil that had made it up this far, that was still threatening this place I loved. It was difficult enough to protect Jamie from what I did know; how could I protect him from what I didn't know?

Another berg rolled nearby. We headed for shore.

The black sand beach was warmed by the sun; my fears dissipated as quickly as the chill in my body. I leaned up against a rock on sand so fine it molded to my shape; I held my baby close and kissed the top of his sweet-smelling head. I could still see about half of Barry Glacier. As Jamie nursed, warm against my breast, a huge column of ice broke loose and fell, like a slice of blue sky, into the water. This was what I had come for; this was what I wanted to share with my new child.

On the return trip I again faced the stern to protect Jamie from the wind, and this time saw the glaciers recede. As the ice field tapered off into a few chunks, Andy called my attention to a sea otter toward shore. It was a female, a mother with a baby on her belly. For a few minutes she regarded us. Then, rather than dive and disturb her pup, she paddled off.

The sun was still high in the sky when we returned to the cabin, but I spent the rest of the day inside with Jamie, napping in the warmth of our sleeping bags. The boat trip had tired us both. Andy left us to follow the coast down to an old gold mine we had both been curious about. I wanted to go explore the gold mine, but I also wanted to stay in the cabin alone with my baby.

"Take lots of pictures," I told Andy as he left. Then I folded my arms around Jamie and drifted off to sleep.

The next morning dawned clear, but a cloud bank was forming in the south. Jamie was asleep, so I walked the beach alone. I felt oddly light without Jamie in my arms; I felt bare, as if I had forgotten something.

The sea otters, twice as many as the previous morning, were again close to shore. Flat water, no waves breaking the silence, the only sound that of the otters eating. Most crunched on crab or sea urchin they had caught in the shallows along the lagoon entrance; one made a *tap tap tap* as it used a rock to break open the shell of its breakfast.

I crouched on the beach and contemplated them. Only one or two nearest me turned to glance in my direction, then paddled a bit farther out.

Another sound broke through the silence—the sound of Jamie crying. The otters dove, and I jumped up and hurried to the cabin to nurse him.

Clouds began to move in, though our cabin was still showered in sunlight. Andy took the boat and explored an old terminal moraine that we had passed on the way to Barry Arm. He was gone for hours. Several times I took Jamie and walked the beach. Clouds began to lower, and soon a fine mist fell, yet I could still make out the moraine and Andy's boat tied to the rocks.

I wondered what he was seeing. I imagined the life that must cling to that long row of rock left behind by a glacier. With so many sea otters around, I knew it must be a rich intertidal and subtidal area.

I wished that I could see it, but I was also glad for this time alone with Jamie. It was one of those times in early motherhood when I felt torn: wanting to go explore as I had before Jamie's birth, and yearning for nothing except to hold him close and care for him; resentful that the bulk of parenting was left to me, and overjoyed that this tiny new human needed me so entirely. To possess and be possessed by another in this way still startled me with its complex power. I had chosen this; I had told Andy it was OK to go without us; I wanted, even craved, this time in the cabin with my baby all to myself.

I was also still tired from having given birth only six weeks earlier; it had been a long and difficult delivery. And I thought I had all the time in the world. Next time we would all go together to the old mine, to the moraine. Next time all of it would still be here, and we would share it together.

Finally Andy returned.

"So, what did you find out there?" I asked him.

"Well, mostly I watched this one immature bald eagle," he said. Then he told me how the eagle, mottled brown and black, stood on the rocks as the tide came in around him. He did not move as the water curled up around his feet, then his body. Finally, when it was up around his neck, he stretched up, shook out his wet feathers, and flew a few feet. Then he swam back to dry land, using his great wings as paddles.

"At first I thought he was injured," he said, "but now I think he was just still learning to fly."

"Wish I could've seen that," I said.

With Jamie asleep, I walked the beach again, from the Port Wells side around to the spit and the lagoon. Clouds blanketed the mountaintops, and a soft rain dimpled the water. A raft of a couple dozen sea otters floated offshore. In the quiet waters of the lagoon, the harbor seal I'd seen every day since our arrival moved silently across water so calm that the mountainside of green, lit by sunlight beneath clouds, was reflected perfectly in it. Only the seal's wake rippling the water gave away the reflection's illusion.

At the end of the gravel spit, I stood still as the incoming tide covered pebbles at my feet. Through the fine mist I saw a white spot on the beach. It was a feather—nearly completely white, with just a few brown speckles, the feather of a young bald eagle. Perhaps it was from the one Andy had seen at the moraine, the eagle just learning to strike out on its own. I picked it up, waved it through the air, felt the lift of flight. I brought the feather, and all it promised, to my son.

The next morning, a light rain fell. We were to be picked up that afternoon, but I was not ready to leave. I bundled Jamie and walked to the beach. Once again, the harbor seal slid across the lagoon, and gulls and eagles fished in the stream. Once more we walked out to the gravel spit, then through the smooth, curved forms of upended tree stumps.

The sea otters were again close to shore. Some turned to glimpse us as we approached, but did not move away. Jamie and I looked on as one dove, reappeared, and brought food to its mouth. It nibbled a bit, then rolled in the water, cleaning the remains from its belly. Two others, fairly young judging by the dark fur of their heads, porpoised through the water, chasing each other. One barked at the other, then made a series of flips in the water, its body forming a circle, head to tail.

Jamie gurgled softly, then made a little bark of his own.

Swim

Jackpot Stream

A hot midsummer trip in southwestern Prince William Sound, everything is lush and green and wildflowered. My son is two, a cherubic toddler with a plump body and never-been-cut blond curls and a constant baby-tooth smile.

We've been out several days, and I want a bath. I want to swim. Here where the water barely breaks fifty degrees in summer, what I call swimming is really wading, ducking, gasping. But that'll do. We've all felt hot and sticky, a rare thing in the Sound.

We leave our boat anchored in Jackpot Bay, and paddle our dinghy up Jackpot Stream. A few fishermen line the banks at the stream mouth, looking for salmon. I want to get beyond them, beyond people. We paddle up and around a bend, and beach the boat just above the people, beside an eddying pool on sun-warmed dark gravel. We pull off rubber boots and socks, grab our towel and Pooh Bear shampoo. We dance upstream like river otters, bounding from rock to rock, splashing from pool to pool. The stream tumbles over smooth, slippery rocks, sounding as joyful as us. We find a gravel eddy to stand in. We strip off sweaty clothes, toss them on the bank, and bathe. I am bathing my baby in the waters of Jackpot Stream in Prince William Sound, and we are giggling as the shocking-cold waters trickle down our skin.

We climb up the bank to dry in the sun, and Jamie runs naked across a meadow, through sedges and daisies and yellow paintbrush flowers as tall as he. I run naked after him, with him. Laughing as he laughs. He finds a fallen tree trunk and crawls up on it. I think about his toes, tender baby feet finding splintered wood, but I won't deny him this barefoot pleasure. It's too rare an experience in Alaska. And it's the most beautiful place I've ever been barefoot, naked, and wet. Lucky boy. Lucky us.

rising tides

1993

ONCE MORE we headed down Passage Canal, out into the Sound, but this time we were in a new boat, a twenty-three-foot fiberglass Bayliner, something safe enough for Jamie. I still wanted to name it the *Glacial Erratic,* but for now it was just called the *o o.* We had scraped off the rest of the letters that spelled *Toy o' Mine* and left just the two *o*s because that was Jamie's first stab at language, something he said when he was excited, or when he dropped something.

We planned to go first to Perry Island, and then on to places we'd seen only on a map. But as we approached Perry Island's West Twin Bay, a queasiness grew in my stomach, a lump in my throat.

This place, here, the year of the spill. A huge black barge had moored in the middle of this bay like a large metal island. Fishing boats, trolling for a different catch that season, hauled oiled seaweed to it; the seaweed was scrubbed cleaned and returned to the water. Two large red *X*s on the side of the barge, orange and white buoys off each of its corners, made it look like something poisonous, something unapproachable, dangerous, a hulking metal mass laying in wait in the midst of the shimmering blue bay.

Only the water lay before me now; only the *o o* anchored in West Twin Bay. Nearly four years had passed since the tanker ruptured on Bligh Reef and spilled oil across Prince William Sound. The first two summers, Exxon had paid for an intensive and invasive cleanup, but now it was over, the barge gone.

We walked the beach at the head of the bay, out to a rocky headland. In 1989, this beach was not listed as "oiled," but I had found many tar balls among the white granite rock. These gooey masses of oil had begun to harden and had collected rock and shell and sand to their surface. They floated on the water like lost beach balls in a game gone bad. Washed up onshore, they cracked open under sunlight and melted into black slime, a mass of stench covering everything in its path.

All that spread before me now were the pale shades of this beach rock and the flotsam of tides: tangled kelp, crab shells, sea lettuce, and popweed in a series of rows like the sea's own garden.

I still looked twice at every dark-colored rock.

In October 1991, when Jamie was three months old, the state and federal governments had settled out of court with Exxon, receiving $1 billion for natural resource damage. For the oiled beaches, the dead and poisoned fish and wildlife. So this was the price of oil; so this was the product of contingent valuation.

From the deck of a cruise ship or the state ferry, the Sound looked clean. But while Exxon and their cadre of scientists had already proclaimed it healed, state and federal government scientists were documenting continued declines in many species, among them Steller sea lions, harbor seals, harlequin ducks, Pacific herring, and pink salmon. These scientists turned over rocks; they dug into beach gravel; they found oil still sheening.

During this time, the David Letterman show aired a spoof of an Exxon ad: there was an image of the Sound on a sunny day, with a voice-over telling us the Sound was now sparkling clean, even cleaner than it had been before the oil spill. Then the final touch: a small star glimmered off a mountain peak, accompanied by a chime, just like the clean teeth in toothpaste commercials.

This joke wasn't far from the mark. In newspapers and magazines around the country, Alaska's tourism marketing associations ran a full-page ad with the famous head shot of Marilyn Monroe, sans the distinctive beauty mark on her upper lip. The copy pointed out that she was still beautiful without it, just as the Sound was still beautiful after the spill.

These attempts at equating the spill with a facial blemish, at denying or making light of the deep-rooted devastation wrought by the oil spill, were more than disturbing; they were making things worse.

◎

By February 1993, Jamie had mastered walking and was learning to run. I had left him with a sitter and attended a science conference in Anchorage held by the Exxon Valdez Oil Spill (EVOS) Trustee Council. This council, composed of federal and state officials, oversaw the spending of settlement funds for restoration of habitat and wildlife. Of the $1 billion settlement, the $100 million in criminal restitution was divided evenly between state and federal governments. The remaining $900 million civil settlement was reduced by two reimbursements: $173 million went to state and federal agencies to reimburse for costs incurred during spill response, damage assessment, and litigation; and $40 million went back to Exxon for cleanup work that took place before the settlement was reached. So far, the rest was being spent primarily on research.

In the large auditorium at the Egan Center, in the dark, a scientist had shown slides and talked about the continued decline of wildlife affected by oil. These were images I'd seen many times before, but they still made me feel ill: that oiled bald eagle carcass, wings stretched out as if in flight; that oiled pigeon guillemot, its yellow eyes staring into the camera the only things not black; that same Sitka black-tailed deer on the beach eating oiled kelp.

These photos appeared over and over on the news, at meetings, in magazines. I didn't want to watch them again, but in the dark auditorium I sensed calm: everyone in this room had seen these images so often that they were no longer shocking; they were in fact what we expected to see.

Another scientist spoke, this time about the human trauma, the social and psychological effects of this human-made disaster. It was all I could do not to leave in the middle of his presentation; I couldn't take any more bad news, and I had been away from Jamie long enough for one day. Time to go home and feed my child.

The talk over, I got up to leave, but from the back of the room, a voice came out of the darkness to ask a different kind of question.

"Given what you've just said about the profound impacts on people that are so intertwined with the ecosystem, don't you think that the most important thing we could do now is to protect that ecosystem from any further damage by buying the coastal forests so they're not logged, by preventing this clear-cutting, which would be like putting salt in a wound?"

I'd heard about this: using restoration money to protect threatened habitat. It was being proposed by a few Cordovans, one of them a marine biologist with the University of Alaska whom I'd interviewed six years earlier for a story on the orcas of Prince William Sound. From the beginning, this man's name had appeared in every major story about the spill. He was always on the front line, always initiating action—in the initial emergency response; in obtaining and placing containment booms; in the commercial fishing community's voluntary cleanup efforts; in the formulation of the Oil Pollution Act of 1990; in the establishment of a citizen's watchdog group, the Prince William Sound Regional Citizen's Advisory Council; and in co-founding the Prince William Sound Science Center.

I found my way to the back where the man sat.

"You're either Rick Steiner or David Grimes," I said to him.

"I'm Rick Steiner," he said. "How'd you know?"

I told him that I'd interviewed him for the orca story many years earlier, that we'd tried to meet a couple of times, but our paths had always just missed. Yes, he remembered. Once, the year of the spill, I had gone out to lunch when he came by. Another time, he missed a focus group I had led for a research project. I told him that I'd followed his work since the oil spill, and admired his efforts, especially his idea to use restoration funds to stop logging in the Sound.

He told me how the settlement had come about—a story that would have been outrageous in normal times. In early 1990, he had been contacted by Frank Iarossi, the president of Exxon Shipping, whom he had met in Valdez just after the spill. Iarossi met with Rick privately in Washington, D.C.—while Rick was touring the country, speaking to groups about the spill—to tell him of a secret criminal plea agreement in the works between Exxon and the U.S. Justice Department. It was a very bad deal, one that would have reduced the U.S. Coast Guard's legal culpability; fined Exxon a ridiculously minimal amount, of which very little would go toward restoration; and delayed the filing of any lawsuit against Exxon for at least four years. Rick called Alaska's attorney general—who knew nothing of it—and, after they broke the story to the *Wall Street Journal* the next day, the plea evaporated within a week.

Having been inserted into these high-level negotiations, Rick realized that he could initiate a different settlement proposal—one that could be used to prevent more damage, especially clear-cutting, to the spill region. Just before the spill, some Alaska Native corporations had started logging the virgin coastal forests they owned within the spill region, seeking to provide income for shareholders. Rick took a one-year sabbatical from the university and, with David Grimes and a handful of others, proposed a comprehensive settlement of all civil and criminal claims between Exxon and both state and federal governments for $2 billion, most of which would go to purchase conservation easements on private lands—to stem this growing tide of clear-cutting. Ever since, Rick had pushed to see that come to pass.

We walked out of the dark auditorium together, out into the sun of midday. We sat outside and talked until I had to leave. I had learned two things that day. I had learned that, measured in long-term effects, the spill was far from over. I had also discovered that at least this one person had not given up, was still working on behalf of Prince William Sound.

@

Dangerous Passage. Icy Bay. Chenega Glacier. Ewan, Jackpot, Granite Bays. One by one, aboard the *o o,* these names on the map became real.

In Granite Bay, a series of rounded white rocks jutted above water like a pod of surfacing beluga whales. At first Andy questioned whether we should take the boat in, but I wanted to try. The sound of a stream wafted toward me and invited exploration. Then he heard the stream, too, and maneuvered to a safe anchor.

Onshore, we hiked a maze of meadows to a small crystal-clear lake. Andy and Jamie waded in, Jamie happy to be free of the diaper. An alarm call, the slapping of wings on water at the far end of the lake, startled us—a pair of loons, one of them rearing up and plowing through the water to ward off us intruders.

On the way back down to the beach, Jamie sat down among sundew and green bog orchids. Below him spread azure waters punctuated with white rock, and beyond him lay all these new places, so much I had yet to see. How wonderful to be in a new part of the Sound with my new boy, to see it for the first time with him, with no memories to overshadow us.

But then the cloud: we hadn't seen a single sea otter since Perry Island. This area had been slammed with oil, and all the sea otters had disappeared. Scientists suspected it would be a long time, if ever, before they returned. What else was missing?

At the shoreline, Jamie busied himself picking up shells and rocks, checking out every cranny. In his tennis shoes and a diaper, he climbed boulders with the ease of a toddler who didn't yet know fear. I stood behind him, hands out to catch him if he should lose his grip and tumble. But he never did.

Halfway back to Whittier, in narrow Culross Passage, a sea otter bobbed ahead of our boat. When we were within a few feet, I held Jamie up and showed it to him. He leaned over in my arms so far I had to hold tighter. He stretched out his chubby arm and offered his partially chewed baby biscuit to this new friend.

Nellie Juan

Derickson Bay

In the foreground are those barren rounded rocks not too long ago ice-covered, and, at the water's edge, two inflatable boats, ours and Peter's, as gray as the rock. Beyond and behind lie the waters scattered with icebergs, and then the massive white striated face of Nellie Juan Glacier. It dwarfs the boats.

Nellie Juan, Nellie Juan. The name sings to me. The place has been called the Yosemite of the North for its rock domes streaked gold and silver by time and minerals. They are massive, steep and rounded, of a different geologic makeup than most of the Sound. But it's not these rocks that bring me back to Nellie Juan.

It's the ice, the glacier, the water.

Crossing wide Port Nellie Juan, first we see white spots on the water toward Derickson Bay. Gradually, those spots become more frequent and larger. Closer, they are ice, at first small, no bigger than my hand, and then increasingly large, as big as our boat. Into Derickson Bay, a long narrow spit of sand reaches nearly across the bay, and beyond it a field of white ice in the sea. The glacier isn't visible yet, just the ice that has calved from it. We ease through the ice floe in our small inflatable, into the heart of Nellie Juan—and there, the glacier. A wall of ice, cloudy white streaked sky blue, landscape of deep crevasses. Ice all around, and on many bergs are the dark, long shapes of harbor seals, dozens of them. In the water, their heads appear for an instant, saucer brown eyes, spy-hopping, then slip under. On that spit, huge bergs are stranded at low tide, some taller than I, each melted into a distinct shape. Silvery blue, bubbles of air captured. An arc of clear ice. Looming ice boulder scored with black gravel.

I wander and take rolls of pictures to send to my sister, a ceramic artist. Come, I want to write to her, come to Alaska and your work can be inspired by ice. You can spend hours on this beach, wandering among the bergs. Wander over to the gray boulders that sand gives way to, and climb these sparsely vegetated rocks up beside the glacier to several ponds, warm enough for swimming on a hot day. Swim in a pond next to a glacier of thousand-year-old ice. Smell the sweet lupine, the fireweed blazing nearby. Hear the cry of kittiwakes over the ice floe, dipping and swaying in front of the glacier's face.

storm warning

1994

I WAS NOT HERE to look for oil. I brought no shovel, overturned no rocks, hunted no pools of sheens or sludge. I knew I could find some—at home I had a jar of *Exxon Valdez* oil collected from a beach two months earlier, more than five years after oil drenched Prince William Sound.

This time I was on Perry Island simply to be where I had been the year of the oil spill. To put myself back into the land alone, to understand what that disaster had invoked in me, feelings that resurfaced last spring as a tidal wave of events that had capsized my life.

All around me was quiet except for the rain steadily beating on my rain suit. Low clouds hid the tallest peak on the island, the one with the glaucous-winged gull colony on top. The only sounds were rain pattering off leaves and waves breaking on the beach. Beyond that beach, whitecaps revealed water too rough for my leaky dinghy.

Five years ago, for the week I camped in this same spot, I had seen no rain. No waves. No clouds. It had been unusually warm and sunny and still. Too warm; the sun softened the oil on the rocks so that black tar stuck to our clothes and acrid fumes stung our eyes. All three of us volunteers, working as oil spill observers, had headaches every day. After that week, we'd moved the entire camp, chagrined at the irony that there was too much oil for the oil spill observers.

Yet this place had haunted me since, as had the spill itself. I had decided to return after the official cleanup was long over and seasons of winter storms and summer births had done their cleansing. I wanted to see for myself how well the place had healed, and to find some way to heal from the disaster myself.

Now I dwelled on the torrent of events that had arisen this spring. Now I sought not only solace but also clarity. Somehow I sensed they were connected—the spill, this place, my conflicting desires. But how?

All winter I had anticipated this anniversary, had felt unsettled, had searched for words to encompass all it meant, but instead of finding meaning, I had found myself reliving it in all its horror. It did not surprise me that I had met Rick again that week, did not surprise me that we had been drawn to spend time together. It did not even surprise me that I found myself in love with him, unable to give him up, even in the face of losing my marriage, my family, and my life as I had known it for fifteen years.

What did surprise me was the turmoil and the fear. And here I faced both: the turmoil of a storm brewing around me, wind whipping up waves, water pouring from the dark sky; and the fear of being alone in the wilderness, on an island twenty-six miles by water from the nearest town. Why had I thrown myself back into this place alone? And why now, when such a storm ranted inside me as well?

But I was not alone. As I set up camp, an eerie yet familiar sound came from overhead, the sound of voices deep in nature. A pair of loons flew across the cove and over my campsite, flying in such synchronicity that the very movement of air from their wing beats seemed to bind them together. They were Pacific loons, with smooth, striped gray-and-white necks, and silver heads set with ruby eyes. I watched their flight, my face turned up into rain, for as long as I could. Long after, stringing tarps to trees for a refuge from rain, I listened to their lament.

The cries of loons had always comforted me, a comfort I now craved, for I was fearful of feelings that had clung to me since my arrival. It was only five hours since my husband and son left me and my gear and my little boat here. Now they were on their way to our house, and I kept wishing I was with them. I missed my boy; this would be the longest we had been apart since his birth three years ago, and his absence was a cold stone growing in my chest.

I was alone, alone with my memories of my time in this place five years earlier. With a half-dozen others, I had come only to bear witness, to share what I saw and didn't see with those who knew only clips replayed on television. Nora, one of hundreds so disturbed by news of the spill that she'd flown from Delaware to help, had hiked up to a lake where a pair of Pacific loons nested. She told me their nest produced no young that year. Loons mated for life and returned to the same lake each summer to nest. Did I just see that same pair? Did they survive the spill years?

A roof over my head, I scouted for rocks and driftwood to create a place to sit and to store my food out of the rain. My campsite was on a short spit of land, with a cove on one side and an open view of the Sound on the other. My kitchen looked out over Day Care Cove.

This was an unofficial name given by a couple who spent many summers kayaking the Sound before the spill. They named it "Day Care" because they always saw so many birds, harbor seals, and sea otters raising their young in this safe, secluded, and nutrient-rich spot. The name no longer seemed to fit; so far, I'd seen only one grebe with her three young.

On the other side of the cove, the land rose steeply. The peak was still shrouded in clouds, but I could trace it in my mind—a granite dome, segmented with crevasses lined with vegetation. At the mountain's base, there was a low pass where a creek met saltwater—the source for my drinking water.

This was among the safest places to camp in Prince William Sound; no giardia in the water and no bears on land—as far as anyone knew. I had plenty of water and food, a dry tent, a handheld radio. I was safe. My fear of being alone was a fear of finding out what I was feeling.

I was in the place I loved, the place that drew me to stay in Alaska one more summer, then one more year. Almost a decade later, Prince William Sound had grown its way into my heart like roots of a tree embedding themselves into soil. It was a beautiful place, a vast and surprising place, but it was not an easy place. To be here was to expect storms; to expect to need winter clothes, a survival suit, a radio to reach the Coast Guard, extra food for being stranded by weather; to expect to be uncomfortable.

The rain poured down, and gray skies enveloped the island. After only a day, I was tired and cold, and my attempts to build a campfire with sodden wood and the few remaining railroad flares that Dan gave us years earlier failed. I fixed a quick dinner, retreated to my tent, and shrank into my sleeping bag.

The wind picked up. It howled against my tent, shaking the walls and beating them with rain and a tether torn loose in the gale. The tent floor billowed up, and I was relieved I'd tied it down with ropes and stakes. Waves broke on the beach, a steady roar punctuated by a thunderous crash every few seconds. The wind made waves crash at a tidemark so high I felt it tugging at the edges of my tent. The rain dashed sideways so that even with my rain fly and big blue tarp overhead, I still had to zip shut the screened window to keep from getting wet.

The roar increased. I imagined crawling out of the tent and being swept away to sea. I worried that my little boat, my lifeline, would be ripped from shore and carried away. Did I tie the line high enough on the beach? Were my knots strong enough to hold?

I ventured out in the slanted light of a summer night. Everything was just as I'd left it, only wetter. The boat was still tethered; my driftwood campsite was intact; the same spruce and alder stood around me; the same boulders lined the shore. The storm was not reshaping the island. It was I who was being reshaped.

For months I had been in a wave, caught up in it, unable to control its direction, its force. I could only hope to ride it rather than drown in it. I could only hope it would not dash me against a rock cliff or throw me, lifeless, onto a beach. The boat that had been my life had capsized, strewing all the contents around me. I didn't know which pieces would sink, which pieces would make it to shore, which pieces would end up with me.

I loved two men. How could that be? Yet it was not that simple. This wasn't about two men; it was about how I had been changing for the past five years. These men were mirrors in which I saw two different versions of me. One showed me who I had been; the other showed me who I could become. One showed me stability and devoted love and a smooth path through a tended meadow; the other showed me passion and amazement and a trail disappearing into thick forest. I had one foot on each path, in each of two very different worlds. I was split in two, camped on an island that was the precise midpoint between where each man lived. How did I choose which to lose?

I sat at this geographic and emotional epicenter and pictured them. Andy in Anchorage, his gentle brown eyes, quiet voice, strong, capable hands, he whom I had known and loved for so long. He was sitting in our house that was, after four years, still in the disarray of remodeling, but was ours nonetheless. Jamie was there with him, my little boy, and as I pictured Jamie a panic arose in me—how could I not choose to be with the father of my child? How could I not choose to keep our small family together, no matter the cost? But I saw Rick, his tall, thin frame; long, beautiful fingers; his soft, passionate voice. Though he lived in Cordova, on the other side of the Sound, he was right now, this week, somewhere in the Sound, flying around and surveying beaches that were still oiled, marking them for a visit by the team of attorneys, jurors, and judge involved in the private class-action suit against Exxon.

Rick was as tormented by the spill as I was, even more so. It was he who had collected the oil in the jar I had at home, he who refused to let people forget what had happened and what was still happening to this place, he who was demanding that the natural resource settlement funds be used not just for research, but for acquiring and protecting habitat.

When I talked with him, whether it was about oil on the beaches or the disappearance of sea lions or the way Jamie tried to feed a cookie to a sea otter, I felt as if the depths of my feelings were finally understood. I felt relief in his presence; I felt as if I was known.

Daylight brought restlessness. Back out into the wet, I walked the beach that fronted the Sound. Waves tumbled fist-sized rocks along the edge of the tide. Beyond this small bay lay the long arc of Lone Island. Thin lines of blue sky revealed snowy peaks of the Chugach Mountains on the mainland—ice fields gleaming white in the sun.

Behind me, the mountain was still shrouded. Maybe it snagged the clouds and held them; they piled up over the cove and kept me in rain. It was sprinkling on me now, but out beyond Lone Island it was sunny.

A shrill cry overhead: two bald eagles sliced the air, curving around the bay to alight on tall spruce across the entrance to the cove. I had seen them there several times, but not once by the eagle nest tree. The nest tree was closer, on a rocky point off this beach. It was easy enough to spot; the trunk was marked with a red triangle.

The year of the spill, a biologist with the U.S. Fish and Wildlife Service had helicoptered to our campsite and hiked over to that tree. I went along. In the nest, he found an unhatched eagle egg, perfect except for one small dark spot. Oil. The parents, he explained, probably picked up crude oil on their feet from combing beaches for food. Then they landed in the nest, and one drop of oil hit their single egg. That one drop killed the embryo.

Most eagle nests in western Prince William Sound had failed that year. This year, I had yet to see an eagle by this one. Instead, crows near the eagle tree chattered noisily and burst from the treetops to circle and chase each other. I suspected the eagles had abandoned it.

For an eagle to abandon a nest was rare. A mated pair built up the same huge platform year after year, using it for decades. Perhaps waiting for an egg that never hatched had made them lose trust in that nest. I knew Prince William Sound itself didn't turn on these creatures; we did. But how could the eagles know that? Did they carry some memory of human cruelty with which to view this event? Did they recall Alaska bounty hunters shooting them out of the air to collect five dollars for every eagle claw? How did the loons, who as a species have survived for ninety million years, comprehend the spill? What in their collective memory of natural disasters, all the violent rendings of the earth, could they compare to this human-made disaster? How could they still trust their chosen lakes with their young?

At one end of the beach was a path through the skunk cabbage and berry bushes to Aleut Beach. The pathway was overgrown; I stepped on the large leaves of skunk cabbage, then up a slippery plank coated with moss. After following more remnants of a makeshift driftwood boardwalk, I stumbled out of the forest onto the beach. A narrow bay, flanked by cliffs, funneled into the short rocky beach, depositing whatever floated by. The white-and-tan boulders of Aleut Beach were coated in oil so thick that the edges of rock disappeared into the ooze. A stench of oil fumes surrounded me. Waves made a muffled sound against the sludge.

I blinked hard; I looked again.

Now the oil was gone, back into memory. The remaining black stains on rocks were lichen, the ooze a woven mass of kelp swept up from last winter's storms.

I sat on a log, heart beating fast. Feeling faint, I put my head between my legs and stared at the smooth log. I remembered it, remembered the stains of oily boot prints on it, made by cleanup workers who had spread oil even as they removed it. Now the log was worn back down to a sun-bleached brown. *Tabula rasa.*

Two oystercatchers flew by low, veering precisely along the water line. Their sharp chatter reminded me that my presence irritated them. I wondered if they were the same pair I had startled here five years ago. Every day I had seen them, no matter which beach or headland I hiked to. I began to think they were following me, but they always acted as if I were following them—staying a few yards in front of me, taking flight and circling around the beach at my least movement, then landing again just beyond me.

I followed the shoreline back to camp and came upon a sea otter bobbing in the surf. His boat-shaped body floated effortlessly as he pounded a rock on the shellfish on his belly. Turning his head, he saw me and dove for cover. Even though I saw him several more times, he was never curious enough to stare back a while before diving.

Before the spill, otters would look at me a long moment, then continue their lolling about or come closer to examine me. Now, otters and seals and sea lions seemed less willing to tolerate a human's presence. Was it that the more cautious animals avoided oil and therefore survived? Or was it that the oil and ensuing cleanup had made these animals less curious and more fearful of us?

I returned to my lonely camp, still unsettled by my apparition of oil on Aleut Beach. I didn't expect the oil spill's hold on me to still be so strong. I'd made nearly a dozen trips in the Sound since the oil spill; I'd even been to this very island; yet this time I was reliving the spill.

This was my first trip alone in the Sound. Except for a few hours now and then on solitary walks, I had always been with another human, usually Andy. Maybe other people acted as a buffer, drowning out the voices in my head.

It was precisely those voices I needed to hear now, for I needed to face all that was surfacing. Whatever it was, it was tenacious. Five years and if anything it had grown stronger. What was I avoiding? I wanted this time in the Sound to help me find some solace, some resolution, clarity, answers. But instead of a relaxing trip, a soothing sun on the beach with which to make peace with the past and make decisions about the future, I got a storm and clouds and rains of discontent.

In the morning, the silver head of a harbor seal trolled the water, turning several times toward me so that I caught her big dark eyes. Suddenly, she disappeared, her black snout the last thing visible. For a moment I was delighted to have spent a few minutes with her. Then I wondered: Was she one of the harbor seals left blinded by oil? Was that why she seemed less concerned over my presence than the sea otter, because those dark pools of eyes were useless?

Night overflowed with dreams. I was with Jamie; the sun was high in a cloudless sky; we were clad in the lightest of clothes. A friend and her son invited us to his party that afternoon, and I told her we would be late because we had another child's birthday party to attend. Then Jamie and I were off to do errands in town. We were both laughing, and I saw his face so clearly: those plump, rosy cheeks; dimpled smile; dark-blue eyes framed by sun-gold hair. A simple day, yet I felt full and happy. Everything was so good, I thought, this must be the day I'll die. Then I awoke to the darkness, alone, rain still pouring down outside the tent, my sleeping bag and clothes damp around me.

Every night, though my dreams were full of people, the only one always with me was Jamie. He was always present, as attached to me now as when he was in my womb. Rick made one brief appearance, laughing, aqua eyes sparkling. Andy never appeared.

Perhaps it was right neither permeated my dreams. For so long, I had not even trusted my own heart—I had second-guessed every feeling, every thought that was at all emotional or intuitive. How could I trust my dreams? Perhaps neither man appeared because I had been asking the wrong question. Not who shall I be with, but who shall I be.

๑

More rain. I felt like a squirrel under this tarp, my midden piling around me: spent matchsticks, coffee bags, tissue paper, pencil shavings, chocolate bar wrappers. I wanted to get out, maybe putter around the shoreline in my little boat or hike some of the hills. But I couldn't find the energy to move far from camp. I was sodden.

The oystercatchers were on the beach again. They crouched low, legs bent and heads down, as if they were stalking something. Their round black bodies were the same size and color as the rocks—only their thin orange beaks and legs gave them away. They remained close together, one always a handbreadth in front. When the leader stopped and sat, the other sat next to it, feathers touching.

Offshore, the grebe swam over to a spot where low tide exposed rock covered with popweed and barnacles. Her three chicks were on the other side of the cove, still within sight, but farther from their mother than I'd seen yet. It was midsummer; they would soon be on their own.

All around were animals mated to place. Grebe chicks letting go of their mother but remaining dependent on the cove; eagles rarely moving their nest site; oystercatchers willing to combat winter's freeze so they didn't have to migrate; loons returning to the same lake year after year; arctic terns flying twenty-five thousand miles round-trip to come back each spring to this one place, Prince William Sound.

We humans, meanwhile, flitted around the globe like bats without echolocation or moths confused by a candle's flame. We were restless, unbounded, unable to commit to a place or to each other. Perhaps I would be better off committing to a place for life rather than a person. Just hunker down on this island, pour myself into this life, and release all the rest that loomed over me like a rain-laden sky.

This was the place that had been the source of inspiration and pain, of solace and sorrow. This was the place I loved more than any other. Why not take my chances here? Why not weather the effects of the oil spill here, on this beach, with these animals whom I cared for? Why not, except for that one child I loved more than anything else?

©

Clouds lifted, rain stopped, and I decided to cross the cove and hike to the lake. The old outboard started on the third pull; the little boat, a yard-sale bargain, moved easily across flat water to the stream on the far side. I tied up and stumbled about on slippery rocks before finding the trailhead, marked by the remains of a small wooden building.

The trail through forest was barely visible, and seemed more like a small animal trail than a human one. Perhaps only river otters had traveled it since the other oil observers and I last walked it.

That summer, we had walked to the lake several times to escape oil fumes. Up in the fields and forests, we could almost forget about the spill. We hiked until sweat ran rivulets down our bodies, then swam in the cold, dark lake, letting freshwater cleanse us. We'd lie on rocks that had never been slick with oil, drying our bodies in the sun. Not until we heard a plane or helicopter overhead, or the steady drone of an outboard in the distance, were we forced back to reality.

Then we'd talk of the starfish convulsing in the poison of a shallow tidepool; the oily footprints of a river otter who had scampered unwittingly through a pool of oil; the fishing boat that came to scoop up floating masses of kelp; the massive barges and dozens of boats and workers in West Twin Bay who cleaned oil off the kelp; the death and destruction caused by oil and the futility and frustration of the cleanup going on all around us, on the shores of Perry Island, on the waters of the Sound.

Now the keening of gulls broke the silence; I was nearing the lake. Pushing through another dense pocket of alder, I came out into a muskeg meadow of sphagnum moss and the miniature beauty of ladies' tresses orchids and round-leaved sundew. I stopped at a heather-covered hummock overlooking the lake. Upon the hummock was a solitary boulder, a glacial erratic left behind long ago by ice, and a single mountain hemlock, two feet high and wind-twisted.

The lake below was immense, several acres surrounded by rock cliffs two to eight feet high. It was so deep the water looked black. On a rocky point that jutted out into the lake was a large Sitka spruce, more than fifty feet tall and as perfectly shaped as a Christmas tree. Its boughs were adorned with a hundred white gulls.

I stretched out on the hummock, next to the tree and against the boulder, hands behind my head. Hearing a sound like breath being expelled, I looked up. A gull flew by, wings carving the air. For a moment, the puffs of air from her wings amplified the gentle breeze rolling across the hummock. She joined a raft of birds in the middle of the lake, cawing and cackling, flapping their wings, splashing water on their backs. I had never understood why the gulls congregated here. There was no food for them; the shellfish and fish they ate were all by the sea. They didn't nest here; they nested up on the mountain. Maybe it was to bathe in the freshwater of the lake.

Well, why was I here? What did I want?

I want to feel the wind on my face in just this way.

I want to feel passion, always. Even if it hurts. Just to know I'm alive and in this world.

I want to know Prince William Sound, all of it. I want to hike over every inch, climb every peak, walk every beach, stand on every headland.

I want to feel this place the way these gulls do—as an extension of myself. I want the borders between me and this place to dissolve in the knowing.

I want to feel more love—pure and simple, not captured in a tangled web. A single, solid love that encompasses all.

I want to fly and sing like these gulls. To hear my wings soar through the atmosphere. To catch the perfect thermal and glide all day.

I want to be fearless.

I want to sit here and watch white birds cluster on a dark tree at water's edge.

I want to trust my heart.

My last day it was cloudy, but no rain had fallen since the night before. I took one last walk through the forest to Aleut Beach and around to the rocky headland of Observation Point. At a bluff overlooking Aleut Beach, the branches of a hemlock perched on the bluff's edge framed the scene—rocky beach, cliffs, ocean, distant islands. It was beautiful. That was all. This was the first time I had seen the beauty of this place not clouded by memory, suspicion, doubt, fear.

Shifting my gaze to the pattern of twisted branches that made up this hemlock, I pulled out my notebook and sketched the tree, every branch, every angle. I tried to be exact, but there were so many twists that my sketch fell off the edge of the page. This tree had led a life textured by wind and rain, contorted into a shape made more beautiful by strife.

How did I choose what to lose? How did this tree surrender the security of a forest with trees all around to live on the edge where wind tests its strength? It had no choice—it came from a seed that was designed to fall away, that fell here, on this cliff edge, carried by the same forces that now gave it shape. Prince William Sound's wildlife didn't choose to lose so much to an oil slick; what happened, happened. The spill had forever altered it, but it had survived. Events were absorbed and became a part of who we were. To resolve to get over them was futile. Life simply went on. Made to endure. Made to persist.

As this tree held through storms—every storm recorded in its every branch, its every bend—so the Sound held the spill. So did I hold my past. I couldn't choose to lose that which was already a part of me. I could only trust my heart as the tree trusted the bluff to hold it, as the loons trusted the lakes they nested upon. They were all acts of faith. All I could do was take my chances with life and hope I weathered them as gracefully as this hemlock.

On my last evening on the island, I sat on an upended stump on the beach. Smooth gray roots curved, fitted to my thighs, and reached to frame a picture of Prince William Sound. Gray-blue waters connected it all: a tan outcropping of rock covered with olive-hued popweed, white barnacles, and black lichen; a rocky headland crowned with tall, dark spruce; a green slice of island floating in watery clouds. All around were voices: gull, crow, eagle, grebe; waves, trees, rock, wind.

Words to a favorite blues song came to me, and I began to sing it aloud. After a few moments, the gulls and crows, eagles and oystercatchers, fell silent. I kept singing. A crow alighted on the top of a spruce near me. I kept singing. I sang an album full of songs, songs of love and loss, the bittersweet blues. All the while, the crow sat and the birds listened. After days of me trudging around silently and listening to their songs, they finally got to hear from me. I added my voice to the songs of this place, and trusted that it would contain them.

Wet

West Twin Bay

In a rain forest, it rains. But the rain doesn't just fall to the ground, soak into soil. It covers everything. It clings so completely that even after the rain stops falling from the sky, even when the sun has shone all day, it still rains. On me. As I walk through it. The trail is spongy with sphagnum moss, and my feet get wet. Even the driftwood planks laid across the muddiest spots are wet, and I slip on mossy wood. Skunk cabbage, as big as my three-year-old son and studded with diamonds of rain, slaps my legs, spraying water. Slender grasses, green orchid, purple aster, and ladies' tresses shiver as I pass, releasing more liquid. This rain is like seed that disperses by attaching to a passing animal's fur, so easily do I gather rain to my body. Droplets seem to jump out onto me, as if dry is simply an abstract concept here, as if an equilibrium of wet on all things must be maintained. Onward. I scramble up a hill, grabbing onto roots. Now salmonberry and blueberry wave wet against my hips. Spruce and hemlock saplings rub my shoulders and pat my back with wet needles. A passing breeze coats my hair and face in the lifeblood of this forest. If I stay here long enough, will the hair on my arms grow moss? Will tendrils of old-man's beard hang around my face? Will my feet sink even deeper into black soil until I am rooted in place?

ashes

1996

REMEMBER: it was in our second summer in Alaska that Andy and I came upon a boat wreck in Prince William Sound.

A wooden pleasure boat lay tipped nearly onto its side at the mouth of East Twin Bay, the engine hanging out of the hull onto a pile of sharp-edged boulders. The insides were gutted, nothing remaining but the saltwater-seized engine and the skeleton of a boat. On the side of the hull, in black letters clinging to peeling white paint, the name was still decipherable: *Robert E. Lee.*

Andy edged our inflatable closer; he wanted to crawl up into the wreck.

I felt something ominous, like a premonition, that its fate was our own, that we would also end up broken on sharp rocks; I pleaded with Andy to back away from it. When he asked why, all I told him was, "I'm afraid it will fall on us."

⑥

I stopped at Mei Mei's office on my way home after teaching. Since meeting in Homer the year of the *Exxon Valdez* oil spill, she and I had stayed in touch, especially since we were now teaching at the same university. We had much in common: teaching, writing, and the oil spill.

"Come in," she said, her arm sweeping wide toward a chair wedged between books and desk. "Can you visit a minute?"

I sat down and felt as I did every time—that she already knew more than I could possibly say. And as always I didn't feel uncomfortable, just relieved.

"So, how's your book going?" I asked. Mei Mei had been working on a novel set during the oil spill. She always seemed agonized when she spoke of it, but now it was finished, and she sounded lighter.

"Oh, Marybeth, would you take a look at it?" she said quickly, leaning toward me. "I need someone to read it who understands what the oil spill was about, to see if that comes across. So, could you read it? I know you know."

In response, I reached into my purse. I pulled out a small glass vial about the size of a lipstick tube, filled with dark-brown sludge. On the side was written, "*Exxon Valdez* crude, Green Island, June 1994." I handed it to her.

"Wow," she said. "Do you walk around with this in your purse all the time?"

"Yeah, ever since Rick gave it to me last summer. Six years, and look at it, Mei Mei, look at what's still on the beaches." I could feel that familiar surge of energy in my voice, something akin to anger but with no target.

The brown sludge was mixed with a few small pebbles, but it was unmistakably oil, fresh-looking crude oil. Not even hardened, cemented, or gluey like crude oil weathered by time. No, this was liquid enough to coat the sides of the glass vial in brown like a thin syrup.

"Oh, Marybeth. Why?"

"I take it out every now and then to show people." And I told her about the previous July, when some friends visiting from North Carolina asked how the Sound was, if the oil was gone. I had pulled out the vial; they were surprised—but I don't know if it was because of the oil or because I carried it with me. "Mostly it's a reminder. A refusal to forget."

"That's what I mean," she sighed. "Who else do I know who'd keep a container of crude oil in her purse? What possesses us to do these things, and how can we possibly explain ourselves to anyone but each other?"

She shook her head. Many of her friends had been urging her to leave the spill behind, worrying that she was stuck in the past. She said she'd even wondered if working on the book kept her stuck.

"But I just can't move on until I understand," she said.

"Yeah," I said. "Shouldn't we learn all we can from this so it never happens again?"

"Well," she said, her voice dropping, "you really ought to figure out what about the oil spill caused you to leave your husband."

There was a long silence, as if she, too, wanted some explanation from me, just like everyone else. Or maybe I was waiting to see what her theory was; maybe she would say something that would make it all make sense, both the spill and my broken marriage. I looked around her office, as if the calendar on the wall, showing an East Coast fall scene, or the picture next to it of her cabin in Homer, held some clues.

"My counselor said I was feeling smothered," I finally said. "And when the oil spill happened, I could relate to it because the literal smothering of animals by oil mirrored the smothering I was feeling in my own life."

"Interesting."

"Yeah, but I don't know if I buy that. I do know there's some meaning tied up in the oil spill that makes it huge for me."

"For me, too. There's a connection I can't explain to other people, and yet it's critical to who I am now."

"That's why it's good to have each other to talk to."

I put the vial back in my purse, and we hugged, wide-armed but briefly.

The year 1989 would always mean one thing to me and to many of my friends. Whenever reference was made to other events of that watershed year—the Tiananmen Square massacre, the dismantling of the Berlin Wall, a birth or a death, a photograph of some family event—we would always say, oh, that happened the year of the oil spill. And we'd find some significance, parallel or ironic, in the association. It seemed like more than coincidence that the same year Richard Heinberg published *Memories and Visions of Paradise,* three articles on the spill were published with the title "Paradise Lost."

But we weren't the only ones seeking some larger meaning in the grounding of the *Exxon Valdez*. The Sound and the spill, place and event, had become sewn together, woven into the cultural fabric so seamlessly that they had become symbols. I saw references to them everywhere. In an article in the *Atlantic.* In action-adventure Hollywood movies like *Water World.* In the TV sitcom *Seinfeld.* In academic papers. Except for us it was all too corporeal to be tamed and relegated to the status of symbol.

As with other benchmarks of time, like the bombing of Pearl Harbor and the assassination of President John F. Kennedy, we remembered the moment we first heard the news. All the events of our lives were now dated prespill or postspill.

Perhaps we were obsessed. Perhaps we were witnesses, repositories of human memory for an event that shouldn't be forgotten. Like the Holocaust, Hiroshima, and the Vietnam War, the oil spill would be remembered because so many survivors would, always and forever, be unable to forget.

In midwinter, Rick moved to Anchorage to escape Cordova and to be with me. Soon after, a friend of his began calling him repeatedly, leaving long and tortured messages. She said no one cared about her, she said Rick's friend who had left her was evil, she said she was going to chop off pieces of her flesh and send them in the mail to all who had disowned her.

She also called Claudia, who had been Rick's girlfriend at the time of the spill. Claudia tried to contact a social worker for her, but, for the sake of her young son, didn't want to take any more calls herself.

"We've got to let her know that the oil spill is over," Claudia told me. "Her problems didn't start there, and she's got to stop blaming it and depending on us to help her."

I wondered what the spill caused and what it only exacerbated. I wondered whether the spill created or magnified this woman's problems. I would never know. That my own marriage ended five years after it—was there a connection, or was I using it as an excuse?

As much as we talked, and cried, and talked, I had not been able to make Andy understand why I was leaving him. I could not explain it to him or to myself. I could only feel it as a need deep in my body.

But I should have been able to give him an explanation. Sobbing, I finally cried out, "I need more passion in my life!"

"OK." He sat back on his heels and said calmly, "That's the first thing you've said that I understand."

But what I meant by passion, I had no idea.

I recalled an afternoon when Rick and I had met for lunch. I was just getting to know him, just getting to learn what this man was about. Before lunch, he had taken a phone call from a newspaper reporter about one of the research projects funded by the EVOS Trustee Council. Scientists had planned to shoot and kill several hundred healthy surf scoters, study their body condition and stomach contents, and put them back in the water to discover where the currents took dead birds. Scoters, one of the most numerous sea ducks in the Sound, had been killed by the thousands; in many places, their population was still declining.

When Rick had heard about this project a week earlier, he was outraged that money designated to help restore the Sound was instead being used to kill healthy birds that had escaped the oil. More than outraged, he was despairing. He had immediately contacted this reporter, hoping she would write an exposé on this inhumane research project that might put a halt to it. But now the reporter had called to tell him that the paper would not be running the story.

All this he told me after he finished the phone call, his voice shaking. Then he wept.

©

Dr. Steven Picou, a psychologist with the University of Southern Alabama, studied the social and psychological effects of the spill, especially in the town of Cordova, where Rick and Claudia and the woman on the phone were all living in 1989. According to Dr. Picou and Dr. Kai Erikson of Yale University, natural disasters are far less damaging socially and psychologically than technological disasters. A natural disaster tends to pull people together—communities and individuals overcome great obstacles to rebuild, and ties are actually strengthened. But a technological disaster like this oil spill tends to pull people apart. It causes division, dissension, and deterioration of communities and individuals. It creates a corrosive community rather than a nurturing community.

In a natural disaster, the cause is a larger force outside ourselves, something outside our control. It is a singular circumscribed event that people can mark and then begin their recovery. In the aftermath of hurricanes, earthquakes, and floods, the stories are of people banding together, of heroism and compassion, of outpouring of aid from far-away places, of unity.

In 1989, when Hurricane Hugo hit North Carolina, it broke all dollar records for damage and was considered the most powerful to strike the United States in twenty years. But when the storm passed, the damage was done; it was finished. Trees could be planted, houses and roads and towns could be repaired and rebuilt. Like the recovery from the 1964 Alaskan earthquake, acts of restoration were clear, and they worked. My sister adopted a dog that had been lost and nearly drowned in the flooding that accompanied the hurricane; Benjamin was now a healthy, happy dog. All this held.

But in a technological disaster with toxic contamination, like Chernobyl, Bhopal, Three Mile Island, and the *Exxon Valdez* oil spill, the bad news just kept coming. Recovery was uncertain and incomplete, fraught with conflict within the community; restoration was indefinable. The oil that killed so many so quickly still lurked beneath rocks, still seeped toxins on outgoing tides. Wildlife still died; fish stocks stumbled; livelihoods teetered. Though theories proliferated, no one really knew what would recover and when, what would deteriorate, what might be happening that we didn't yet know about. No one knew when, or if, the oil spill would be over.

This chronic aspect was compounded by the stark knowledge that it was our own fault. Directly or indirectly, we caused technological disasters to happen to each other. Whether it was the humans who owned the oil, who owned the ship, the human piloting it, the humans tracking it, those making laws and regulations, or those driving cars and using oil in thousands of ways every day—we were all agents of the disaster.

We were both victim and criminal. To hold two such opposing views required a larger container than most of us had. It spilled out; it flooded into the family, the home, the community, the world. Everyone was to blame, and no one was to blame.

In the aftermath, especially in spill communities, there were suicides, breakups, lost friendships, and exoduses. Within two years of the spill, a man who had lived on the outskirts of town for more than a decade, and who was a kind of "seer" to many residents, committed hari-kari. Shortly thereafter, a lifelong resident and mayor of the town committed suicide. Neither Rick nor Claudia nor the woman who called lived in Cordova anymore. There were still hard feelings between previously close-knit groups over who went to work for Exxon and who didn't, over who got paid big bucks for renting their boats and who was overlooked or chose not to "sell out."

Natives living in the spill region's communities no longer trusted the beaches and water that they had depended upon for thousands of years. Their increasingly rare place in the world—as indigenous people who relied upon hunting and fishing—was at risk. With their mixed cash and subsistence economy, they began purchasing more food and taking less from the land around them—deer, seal, salmon, shellfish, kelp—this medley of food that maintained not just their health, but, more important, their cultural identity and community integrity. They mistrusted officials, too, who gave them reports on which beaches were safe and which were still contaminated. Their lives upended, the population of the Native village of Chenega Bay dropped from one hundred to thirty-five.

But who knows what caused what. Perhaps the weaknesses were there all along, and the spill had brought them to the surface. Alaska Native cultures had suffered other setbacks; suicide rates in Alaska were chronically higher than the national average; Alaska's population was renowned for its transience. For years, Andy and I had recognized problems in our marriage; we had even sought counseling a couple of times. But we were busy, and we always thought we had plenty of time to iron out our differences. Maybe it was coincidence, the oil spill and the breakup of my marriage. But that's not how it felt. I could trace back quite clearly the path of an uneasiness, trace it right back to that shattering event when something I had taken for granted blew apart.

And not until I began living apart from Andy did I learn where we ended and I began. We had been together for so many years, shared so much growing together—my last years in college, our marriage, our land, our cross-country trip that landed us here, the birth of our son—that I could no longer recognize how entwined we had become. Not until I began to unravel the tight weave of our life—thread by thread, each time a shock to see that this thread wasn't me, it was us, this was one more thread I was losing—could I begin to know the full magnitude of this breakup, could I begin to see that it would take a long time to reweave the threads that remained into a new cloth, someday a whole cloth.

I had taken for granted my marriage; I had taken for granted Prince William Sound.

Many who lived along the edges of the Sound, many who fished its waters for their livelihood, who collected mussels and clams from its beaches, also took what they had for granted. They never saw the spill coming, either. An entire region of people was hit broadside, laid flat, left bewildered.

A few knew that an oil spill was inevitable, but they had been considered pessimists, naysayers. These were the ones who had lobbied hard for the Canadian overland route for the pipeline, precisely over fear of what now had come to pass. Overnight, they became experts whose words many of the bewildered clung to for hope or used to affix blame and exact revenge.

As a commercial fisherman and the marine advisory agent for Prince William Sound—a scientist whose knowledge was trued by experience—Rick was one to whom people began to listen. He was also one they turned to, and then, in some instances, turned on. He became a human lightning rod.

In a moment of portent, one person can rise to leadership simply by being willing to take action. Later, when the sense of immediacy has left and people have returned to their private interests, that person is often one of the first to be stoned. So it was for Rick. His efforts at bringing about the quick settlement were applauded, but when the money came through and he advanced the use of it for habitat acquisition, many in the community disagreed—and let him know, loudly and sometimes threateningly. By the time he finally moved from Cordova, he had nearly completely isolated himself from the community.

The oil spill caused collateral damage in the same way that an accidental death could destroy a family. It was the kind of disaster that made people lost to each other, unable to reconnect no matter how much they might want to. I felt that way about my marriage. So many times over so many years I just wanted us to go back to the way things were, to before we split up, only to feel this great expanse between us, this bottomless chasm.

I carried that vial of oil with me for almost two years, loaning it at one point to Naomi so that her ten-year-old son, Nolan, could use it in a science fair. Nolan had been only three years old when the oil spill hit; now he was on the edge of adolescence, beginning to make his own judgments about the world around him. An entire generation would soon grow up in the wake of the spill. The oil spill was passing into history, a single event laid alongside other disasters in history books, not an ongoing problem that still caused harm. My own son would know of this event only through the stories of his parents, the way my generation knew about the Great Depression and World War II.

I wanted Nolan's classmates to see the vial of oil; I wanted them to ask their parents about it, to remember and to remind others; I wanted them to do better than us.

During the few weeks that Nolan had the vial of oil, I awoke from a dream in which I was camped alone on the shores of Harriman Fjord. In the dream I was lying in my tent, the slant of morning light warming me, birdsong awakening me. The hummingbird's thrum, clear flute of the thrush, a tern's rough trill.

It was a brief dream, but so vivid that in my first breath upon awakening, I could almost smell the sharp scent of crushed spruce needles and the sweetness of heather in bloom. Brief, but strong.

I had needed this dream. I had been going to meetings, writing letters, giving testimony on behalf of Prince William Sound. But I had been spending less time actually in the Sound. I felt as if I was losing touch with the very font of my caring, of my actions. This was the irony of activism: it took place high in skyscrapers deep within cities, in windowless rooms of beige and brown, where the only sounds were the humming of computers and whirs of fans, the only scents that of fresh ink and aftershave. This was the curse of activism: it took place far from the source of the passion, far from the living, breathing world that was both its genesis and its sustenance.

௵

The next fall, Mei Mei moved to Seattle and entered a Ph.D. program. Her novel had not yet found a publisher, but she had set it aside.

"I've been torturing myself with that novel for six summers," she told me. "I decided this summer I wanted to enjoy life."

"Well," I said, "with something as huge as the spill, we want terribly to understand it. But just because we want to doesn't mean we get to. Sometimes, at some point we just have to move forward into our lives anyhow. We can only trust that someday we'll live into an understanding."

"Oh," she said. "That's beautiful, Marybeth."

I was glad that my words sounded true to her, that she was now feeling liberated after so many hard years, "uphill years" she called them.

But I wasn't sure I believed my own words. I was parroting the poet Rainer Maria Rilke, a quote I had carried around with me for at least twenty years: "Be patient toward all that is unsolved in your heart. . . . Try to love the questions themselves. . . . Do not now seek the answers, which cannot be given because you would not be able to live them. And the point is, to live everything. Live the questions now. Perhaps you will then gradually, without noticing it, live along some distant day into the answers." Oh, how I wanted to believe his words. How I wanted to live along into the answers. So far, though, all I had to show for it were more questions.

I had struggled to comprehend the spill's effect, agonizing over it, pushing my brain for an answer like Sisyphus pushing the rock, and to no avail. I had watched others do it, too, trying to come to resolution without forgetting, without denying. When was it "over," as Claudia had said? Where was resolution?

Some said it would be over after the civil suit was paid out. The economic reverberations of the spill were great, especially in commercial fishing. The value of permits plunged to a tenth of their prespill value as fish populations crashed. Rick and a friend had bought a salmon seine permit one month before the spill for three hundred thousand dollars; five years after the spill, they cut their losses and sold it for seventy thousand dollars. Now, that permit was worth less than thirty thousand dollars. Commercial fishermen, spill-region residents and businesses, and Native subsistence communities had filed a class-action suit against Exxon for economic damages as well as pain and suffering. Some believed that once that was settled, people could take the money and repair their lives.

Some hoped to find resolution when the forests of Prince William Sound were protected—when we stopped doing more harm to the place. Five years after the settlement, much of the Sound's forests were still at risk from clear-cutting. The spill would be over for him, Rick said, once settlement funds had protected these privately owned forests.

Still others hoped to find resolution by moving out of the state, by starting new lives in new communities with new friends who wouldn't know what they meant when they said "oil spill." There were times I longed to take this option.

Instead, I planted a garden in Anchorage. I was renting an old 1960s mobile home perched high in the Chugach Mountains ringing Anchorage. I told myself I could leave anytime. Everything, I told myself, was temporary.

But it wasn't true. I was rooted. In motherhood, in Alaska. When my Grandfather Bruno died at the age of ninety-one, I planted in his memory. I wrapped my fingers around the stick he'd carved from the limb of an oak, the one he used for making seedling holes, my fingers on wood worn smooth by his rough hands. I planted basil, and oregano, and Italian parsley in a terra-cotta pot made in Italy. Like him. Planted a tomato—Siberia was its name; the oxhearts would never have made it this far north—in a black plastic five-gallon bucket, shielded it with a clear plastic dome, working hard for a few little red fruits. Planted zucchini in mounds, and spinach in rows. I dug deep into dark soil, the glacial till, the humus of a forest, digging, digging, for some deep answer to my longing, some kind of resolution I could hold in my arms like a harvest of sweet corn, like my sleeping child.

Meanwhile the Chugach Mountains stretched behind me, all the way on their steep, spiring march to Prince William Sound. Merlins came to nest in the spruce, and at night a great horned owl came for the community of shrews living off spilled birdseed.

I did not know how to move forward into my own life. I planted, I prayed, I pushed for some knowing. I struggled to make sense of my divorce, to understand what had happened, to have some way of explaining it to others, to someday be able to explain it to my son, Jamie.

I felt horrible about the way I'd left my marriage. I worried constantly about how the divorce and joint custody were affecting my son. I hated shuttling Jamie between two houses, even though Andy and I worked hard to make it as smooth for our son as possible. Jamie was only three when we separated; I knew nothing about a child's life, about how this would affect him when he was five, or ten, or sixteen. Nothing.

On the long nights he stayed with his dad, he and I talked on the phone. And sometimes, he cried, sobbed, into the phone, "I want to be with you. Why can't I be with you?" And I sat alone in that tiny metal house, hating myself, hating what I had done, was doing, to my child. Wondering, every time, whether any of it needed to happen.

I knew I had to move forward, regardless of how little I understood. All I knew to do for Jamie was to croon into the phone, "It's OK. I love you. I will see you tomorrow, in just a few hours. Go to sleep, and then when you wake up it will be sooner. Now let's hug each other tight." And then I'd hug myself as he hugged himself, each of us feeling loving arms around us. All that was left was to start from wherever I found myself, whatever beach I had landed on with whatever pieces I had left, to recite Rilke's quote over and over, and maybe start believing it.

Outside in the light of an Alaska summer night, the tiny buds of basil released their pungent scent, and zucchini unfurled broad yellow blooms. I wanted to trust in myself again. I wanted, more than anything, to just let go.

On the night of the seventh anniversary, Jamie and I walked out into our yard, the lights of the city fanning out below us in the dark. We dug a shallow hole in the winter's accumulation of snow. I lined the hole with foil, and Jamie helped me pile up a few dry twigs and crumpled paper. On top I laid the vial of crude oil.

I lit the paper, and my son and I watched it burn. At first it burned slowly, and I thought the fire might go out. But the glass shattered under the heat, the oil caught fire, and the flame burned with renewed intensity. Tall orange flames jumped above our heads and into the night sky, illuminating the new-fallen layer of snow, a young spruce, and my four year old's face. For a moment, the flames were so high that I became afraid and made Jamie step back. But we stood upon several feet of snow; the fire had nowhere to go except up. The flames leaped higher and higher, blasts of energy released into the vast darkness. Then, just as quickly as they had started, the flames lowered and shriveled the remaining paper, succumbing to the cold all around. All that was left was the small black lid, which held my gaze too long.

Observation Point

Perry Island

Again I walk the narrow trail through skunk cabbage swamp, up Sitka-spruce ridge, and out onto bare rock, climbing to the highest perch on Observation Point. It's another clear day, and from here I can see out to the Dutch Islands, to Lone Island, and beyond to Naked Island, where the Exxon Valdez tanker is now, four months after it hit Bligh Reef, crippled and awaiting a tow to a southern harbor for repairs. The oil has long since spread farther.

It's about noon; I've come to this point every day for four days at this same time, to be consistent in my role as observer for this volunteer project. Some baseline marker for the details that will follow. Some solid reality from which to work. Something I can count on.

I sit cross-legged on the rock, get out my water, snack, waterproof pad in which I record each moment. "11:55 a.m., Tuesday, July 2, 1989, Observation Point," I write. I scan the beaches and see two oystercatchers walking on Aleut Beach. I name some of the wildflowers around me: "yellow monkey flower, white saxifrage, purple butterwort." Just as I begin to lift my head to scan the water, I hear a sound of breath and water expelled, and look out.

There's a Steller sea lion, arcing through the water just offshore from where I sit. It's a small one from the looks of its golden brown head and back. The long whiskers sparkle in sunlight, the skin glistens like water. It moves purposefully, as it has every time before, steadily and in a straight line, a graceful undulation of sleek brown body through a mirrored surface. Only once does it turn its head toward me, and that without changing the pace of its smooth porpoising.

I've seen a sea lion pass this way, north to south, for four days in a row now, same direction, same time of day. It's probably the same sea lion. There's always only one, though they usually travel in groups. To the north there's a headland with a cluster of smooth tan rocks just above the high-tide line. I've been told these rocks are a favorite haulout for Steller sea lions, but the two times I've gone there to look for them, I've not seen a single one.

Just this one. And suddenly I realize that I am observing a wild animal's routine. I've never considered that just as we have our routines, so must wild animals. That they may feed, haul out, congregate, travel in the same places at the same time day after day. I'm comforted to find this bit of similarity, to see this bit of normalcy, this animal going about its days in a routine that, I hope, has preceded and will outlast this terrible time.

PART III

Wisdom is always motivated by love. And love is . . . defined as much by what it doesn't do and will not do as by what it does.

DAVID ORR, in *Listening to the Land*

The beauty of things was born before eyes and
 sufficient to itself, the heart-breaking beauty
will remain when there is no heart left to break for it.

ROBINSON JEFFERS, "Credo"

Fireweed

Alaganik Slough

Beside me on the banks of this sluggish gray river is a flaming field of dwarf fireweed. Now I see why it's also called river beauty. I open my drawing pad to a clean page, pull out my pencils, and try once more to capture the exact shape of leaves and petals, the precise color of them.

Fireweed. The two species, dwarf and tall, are the delicate beauty arising from powerful elements that define Alaska. Fragile and intricate, they are among the first to appear on disturbed ground, the first on charred soil after a fire, the first on scarred ground after logging, the first seed-bearing plant on barren rock just released from the icy tongue of a retreating glacier. That's where I saw fireweed for the first time: beside Nellie Juan Glacier, rimming the rock with its fire. Now I etch the oval leaves in gray and green.

Fireweed. I'm puzzled by its name. Fire: with weedlike tenacity, it ignites a fire of green on barren ground. Fire: it blazes across mountainsides in mid- and late summer, blooming profusely and then bursting white cotton-swathed seedpods that look like wisps of smoke as they quiver and release to the winds. How to draw such ephemeral things?

Fireweed tells me how far along summer is. Once it starts blooming, summer is halfway through. Blossoms start at the bottom of a long spike stem, buds drooping as they elongate, opening into four large petals, a cluster of yellow stamens and pistil in the center. Once the blossoms are farther up, and pale seedpods begin to form below, I start thinking of winter. The released seeds, wafting to the ground, foreshadow snow. I lower my pad and lean into one flower, dusting fiery petals with golden pollen.

Fireweed: why the weed? A weed is a plant growing in the wrong place. How could fireweed ever be growing in the wrong place? It is out there alone on barren rock, blazing a trail for those who come after it: alder, birch, spruce. The trail of succession, the return of forest, phoenix rising.

bull's-eye

1996

ON THE BEACH at the tide's edge were eleven oystercatchers. Until then, the most I'd seen together on a single beach was three—an adult pair and their fledgling. As I approached, their chatter became louder and louder until they all rose together in an arching flight. Eleven round black bodies, eleven thin orange beaks, a line of eleven pairs of black wings flying low out over the water and then back to the beach, landing in front of me.

I was glad to see oystercatchers; I thought of them the way scientists think of an indicator species. They were like the running cedar I'd look for in the woods of North Carolina: if the forest floor was threaded with the dark-green vine with its flat fans of needles, the forest was healthy. Running cedar grew only where the air was clean; oystercatchers lived only where the beaches were undeveloped, the land undisturbed. Their presence on this beach was a sign that all was well.

But to see this many together was new. I watched their meandering tide walk, the way they synchronized movement with the waves. Then I turned and walked in the opposite direction, down the long, flat beach flanked by rocky headlands, bounded by alder thickets and steep mountains and a flat expanse of blue-green water studded with ice.

I dropped into my beach trance, sounds and images drawing in and out like breath, time and place melding. Surf washed up, pulled back, water pushing cobblestones over each other, rubbing, polishing. From behind came the sharp crescendo of oystercatchers, far out of reach.

Back at the cabin I was sharing with my friend Gretchen, I told her about the oystercatchers. This was one of only a few times I'd been out in the Sound without Andy. I enjoyed sharing the Sound with a new person, but sometimes I missed the way shared memories could deepen a new experience.

"Do they eat oysters?" she asked.

I wasn't sure.

"Well, they eat mussels, clams, other shellfish, too, I think," I said, recalling that there weren't, as far as I knew, any wild oysters in Alaska. Gretchen had lived in Alaska only two years; she looked to me to know this place, to know what oystercatchers ate.

I'd watched them at Perry Island as they picked among cobblestones at low tide; at Surprise Glacier, as they stood at a headland, three orange beaks pointed into the wind. They would often walk in front of me, silent, crouched low, bobbing their heads as if they were muttering to each other about me, then start their high-pitched cry, and fly out over the water and back around to a point farther up the beach. They'd repeat this crescent-shaped flight several times until they flew around and landed behind me to continue their patrol, this time following me.

They seemed agitated by my presence, but rarely left entirely. They weren't shy, especially when protecting their young. Several times I'd come upon what must have been a nest site. The adults would fly erratically toward me, land, drag one wing, cry endlessly in a high-pitched whistle, frantic in their attempts to lure me away from their egg or young chick. That's something I liked most about oystercatchers: they didn't just fly off at my approach; they stood their ground.

But what did I really know about black oystercatchers? I consulted the guidebooks we'd brought. *Haematopus bachmani*. The only member of the family *Haematopodidae* in Alaska. Of an estimated fifteen thousand worldwide, half live in Alaska, and 20 percent of those live here, in Prince William Sound. One book described them as "large, dumpy, short-legged" shorebirds. I'd never thought of them as dumpy; I would have described them as compact, round, sleek as the rounded stones on the beaches they walk.

They push the chisel tip of their beaks between two shells of a mollusk to pry it open. A perfection of form and function, that beautiful fit that characterizes the natural world. They probe in the sand and rocks for marine worms. Pry limpets and snails from rocks. And they do gather in large groups; "family groups" is what one book called the gathering I saw. In winter, they don't migrate south; they gather together in larger flocks, as many as a hundred.

I imagined that sight: oystercatchers on the beach. One hundred of them. A sea of black punctuated by thin orange lines and pairs of those eyes: absolutely round black centers, yellow irises, and orange eyelids, upper and lower—perfect concentric circles. I'd like to see that, a flock of one hundred oystercatchers.

@

I enjoyed learning what science had uncovered about an animal or place, especially through prolonged observation, such as Adolph Murie's wolf studies in Denali National Park. When I saw the first arctic tern of the summer, I liked knowing that this bird that migrated the longest distance was also one of the first to return. I'd been a student of the biological sciences; I'd worked on a variety of research projects. I'd spent time with biologists studying trumpeter swans; we sat in camouflage-colored tents hidden among the trees, observing a family of swans; we recorded the amount of time they spent feeding, preening, sitting, swimming, flying.

But science gave me only one kind of knowledge about swans and oystercatchers, a knowledge that had its origins in Descartes's dualistic view of the world: everything was either matter or spirit, and only humans, with their God and their ability to reason, were both. To Descartes and his fellow philosophers of the Age of Reason, everything in the universe behaved according to a few simple mathematic laws—an oystercatcher was no more than a box of discernable parts that, when put together in a certain way, would perform like an oystercatcher.

I wanted more than a one-dimensional puzzle, a one-way river of information, more than this process of breaking down the lives of swans and oystercatchers into manageable pieces. Such breaking down into pieces, a Zen Buddhist master once said, actually separates worlds rather than bringing them together.

Relating brought worlds together. And that's what I wanted. In relating, I wasn't in the kind of hurry that a scientist might be; I didn't have those kinds of targeted questions; I didn't need to fit my findings into a particular discipline. For relating, that give-and-take I sought, I was better off spending time with oystercatchers on a beach.

And yet, when I had come upon eleven oystercatchers, I had turned around so that I wouldn't bother them anymore. I had left them alone.

It was lonely being human. When I saw wild animals, I wished they wouldn't run or fly or swim off at my appearance; I wished they wouldn't be so frightened. I wished they enjoyed my presence the way I enjoyed theirs.

As a child, sometimes I'd imagine myself a seabird, or a dolphin, or a wolf. I learned early that they faced many dangers in the wild, not just from humans but from each other. If I were a seabird chick sitting on an exposed nest, a bald eagle overhead would be cause not for wonder, but for terror. But that community of animals, that perfect fit within the land—I yearned for it.

Among the remains of my Catholic upbringing was a picture of paradise: the lion and the lamb, the man and woman, the birds and beasts all lying down together like the greatest of friends. There was some sense, even in this belief system where humans have dominion over earth, that relating to other animals was a blessing, was, in essence, the direction of all our yearning. I knew few people who expressed a desire for such connection. But to me a life fully lived required a deepening of the bonds between humans and the more-than-human world. This was more than a solution to the ills befallen our planet; it was our birthright.

Relate: To interact. To respond. To have relationship or connection. I wanted not just to know about oystercatchers, to identify them and know their scientific name, their habits and habitats. I wanted to experience something shared, some sense of reciprocity, of trust.

&

My home in the little trailer hugged the Chugach Mountains so tightly that my yard was a wildlife corridor. In the springtime, especially, young brown and black bears would wander through, just awakening from their long sleep, hungry and searching for a territory to call their own. Beside the front door to the mobile home, the metal siding was scratched and dented by a brown bear who had visited years before I moved in.

One afternoon in late May, I walked out onto the deck to check on my newly planted garden. A few thin clouds streaked the sky; the elderberry bushes lining the deck carried the slightly acrid scent of their white blossoms, as if to forewarn of their poisonous leaves and stems.

I turned the corner and came face-to-face with a small black bear. She was standing on all fours on the deck, her nose just inches from the herb planter. We stared at each other full-on for one long moment. Caught in her gaze I forgot to be afraid. She jumped off the porch, right into my garden, stopped and turned to glance at me again; I stood still, facing her; she spun around and galloped through the elderberry and down the hill.

In between gangly seedlings was one perfect bear pawprint. I left it there all summer, planting and weeding and watering around it. Sometimes it filled with rainwater and glimmered in the slanted light like a small pond.

Maggie, as we came to call this bear, visited so many times that we began to expect her. Once she ambled up to the window where Jamie and Rick and I stood, watching and containing cries of delight. But in midsummer the visits ended. After that, we could only wonder if she was the black bear that was caught and radio-collared in the park below us, if she was the black bear shot and killed by our neighbor for bothering his dogs and taking one of his chickens. He didn't want her skin because he already had three brown bear skins in his house, and that was all he had room for.

My first few years in the Sound, I felt as if things were in proper scale. The place seemed vast and wild enough to contain a few humans without the other animals being run off or killed off. After growing up on the crowded East Coast with its beleaguered wildlife, I was relieved to feel neither threatening nor dominant in the presence of wild creatures. These feelings returned, though, after the *Exxon Valdez* oil spill and cleanup. After all that the wild creatures experienced, I sensed that the surviving animals were so traumatized that they were more wary of me now. I feared a certain innocence in my relations with them had been lost, or at least severely crippled.

Harbor seals were less curious than before. Rather than following and spy-hopping, the ones I'd seen since the spill dove quickly and swam far away before resurfacing. As the poet Jane Kenyon might have put it, they were "surfacing far away from humankind." They weren't so interested in us anymore. I wasn't the only one to notice this heightened wariness—friends who studied sea otters told me the otters dove sooner, that they weren't able to get as close as they could before the spill.

I felt like an intruder, a pest, a danger. I'd see a pair of marbled murrelets floating on the water, remember that their habitat was disappearing along with the old-growth forest, and steer in a wide arc around them. I'd see a harbor seal sunning on an iceberg, recall how their numbers were plummeting, and turn the boat around so that I wouldn't cause it to dive into the water. I turned away from them, again and again, because I didn't want to make things worse.

Oystercatchers, spending all their time along shorelines, were particularly susceptible to disturbance. According to a biologist who spent time observing them, oystercatchers suffered in the Sound not so much because of oil, but because of the massive disturbance of the cleanup efforts.

Long after I had left sea otter number 73 in her cage in Seward, long after all the rescued sea otters had been either returned to the wild or sent to a zoo, I wondered about number 73. I had run my hands over her wet body, held her head in my two hands, learned every inch of her physical being. She had been unconscious or so heavily drugged and disoriented that I doubted she had any memory of me. But maybe she did. Maybe she remembered human hands and human voices. If she had survived to be returned to the wild, and if she remembered her human captors at all, I hoped that the memories were not all bad.

I regretted what we put her through, not just the oil itself, but the capture, drugging, tagging, cleaning, and caging. So what if I desired some reciprocal relation with wild animals; what was it worth to them? What could oystercatchers get out of my being on their beach, except annoyance and maybe even terror?

Older memories held me back as well. I could still remember, with fresh anguish, a time when I was six years old, taking care of my sister, who was barely two. A warm afternoon, just after Easter, we were sitting in the yard and playing with our new duckling. It was a yellow duckling, not one that had been dyed blue or pink or green. Just bright yellow. My sister was holding the duckling and laughing, her yellow curls bouncing over her shoulders. She held up the duckling by the neck, and I started to tell her not to hold it so tightly, but then I saw the duckling's eyes bulge, its head droop. It was dead; we had killed it. My sister didn't understand dead, but I did. We buried it in a shoe box in the backyard, next to the rabbit that our dog, Pepi, had dragged out of a nest the previous year. Our baby animal graveyard.

Animals, according to some scientists, have a collective memory, a species memory. Whales remember harpoons; mountain lions and wolves remember poisoned carcasses; seals remember their rookeries being bombed for bounty. They adapt, they evolve, they survive. If a virus could develop immunity to a vaccine, then couldn't sea otters recall the capture of so many of their kind that they learned to dive at the first awareness of a boat or plane?

All this made me less willing to risk disturbing them for the slim chance of resurrecting a connection, a moment of mutual awareness. Maybe there was nothing mutual about it anymore. Maybe they had just had enough of us.

And yet—when the oystercatchers landed in front of or behind me, rather than flying far away from me, I couldn't help but wonder if they were curious about me. Maybe that was where relating started: in simple curiosity.

©

Hundreds of ravens spent winter days in the city, scrounging French fries and stale bread from Dumpsters, returning at dusk to roost high in mountain forests. Near our trailer at the foot of the Chugach, I'd see them flying overhead, always around the same time every day. I began timing my walks to be in an open meadow when they flew home; I counted as many as fifty overhead at one time.

Small wonder ravens are prominent in the mythology of Northwest Coast Natives. Curious tricksters, like the coyote of the Southwest, they were good choices for practice in relating. I began doing more than just standing in the meadow, looking up, agog. Sometimes I danced, waved my arms, sang a little. Once I even brought a shiny piece of metal—I had heard they are attracted to things that reflect sunlight. Several winters earlier, a growing number of people in Soldotna, a small town on the Kenai Peninsula, had reported stolen windshield wipers; everyone had been stumped until a raven was caught in the act. So I waved a shiny silver strip over my head; a few ravens circled back around for another look.

Sometimes I took Jamie on these walks. At six, he thought this was a great game; how fun to watch Mom be so silly. And how much easier for me to be silly when he was near. I glanced at my son, noticed his movements—slow, aimless, mindless. A rock might catch his eye; he'd pick it up, drop it, pick up a stick. He'd sweep it before him as he moved through the tall grass of midsummer, letting the blades brush his arms. He moved more like a wild animal, more like the ravens, in tune with his surroundings. He was teaching me how to relate.

Jamie alternately communicated with the ravens and explored the meadow; he wasn't so single-minded as I. And he showed none of my hesitation or doubt. Of course he could talk to ravens; of course they were talking back. He conversed with the blades of grass as well.

On one walk back home, after the ravens had headed home themselves, Jamie asked, "Who do you think God is, Mama?" We'd already had a few talks about God; I had told him that people believed different things, that I believed God was a spirit that was everywhere, in every thing, every rock, every raven. But this time, I asked him, "Who do you think God is?"

"I think he is this head that floats around everywhere, and you can't see him but he's there."

Like the language we shared with ravens, invisible but ever present.

Jamie mimicked raven calls, mastering them better than I. He would gaze up at the black crescents of ravens overhead, and trill a resounding "caw, caw," then summon up a guttural croak. Once, several ravens circled back, swooping lower and lower, the furling and unfurling wings dropping closer, until we could see them tilt their heads at us. Tighter and tighter, lower and lower, within a few feet of us. One gave a single caw, and Jamie cawed back. They glided above us, then lifted, toward the mountains again. Another time, five followed us back down the trail toward home, flying closely overhead as Jamie kept up a steady conversation with them.

But these were city ravens, bold and curious by nature and nurture, and they were far from putting themselves at any risk with us. Like magpies, jays, gulls, and squirrels, ravens tolerate, even benefit from, humans. In a creation story of the Northwest Coast Natives, it was Raven who pried open a clamshell on the beach. As it opened, people came spilling out. The Pandora's box of the wild world opened by Raven.

Perhaps ravens are like ambassadors from other countries, emissaries from afar. Buffer for and bridge to the wild, protector and translator—the way newborn babies are considered emissaries from the spirit world in Tibet, their sacredness kept intact, the celebration of their first year on Earth marked by touching their feet to the ground for the first time.

The oystercatchers, though, encountered me on their home beaches, near their nests, not after a day in the city. Their tolerance was not that of ravens. But maybe they, too, were ambassadors. The way they stood their ground, as if to see whether I could find that balance, whether I could put myself in proper scale with them and their beach.

If the oystercatchers could experience a human doing something that didn't feel threatening, but was just interesting, that might be a good thing. Not just for me, but for them as well. For their species memory.

⑥

I went back out on the beach. They were still there, the eleven oystercatchers. One thing I knew about them held for now: they had a very specific and relatively small territory. I sidled forward, not looking directly at them. I remembered the bird-rescue class in Homer: a direct gaze signals predatory intentions to birds. A few yards away, I crouched down on the beach. They didn't fly off, but they did whistle, a quiet, breathy sound. One started to fly, wings lifted out and up, then only hopped a few steps and folded its wings. They resumed pecking among pebbles at the tideline and wandered my way.

A wave moved a few stones near me. I reached out one hand and poked at a few pebbles, flipping one over. It clattered, tumbled. I noticed the multitude of colors, greens and reds and blues, that the bathing by waves had revealed. I began to hum softly. The oystercatchers kept advancing toward me; one looked right at me. Then another looked. Those brilliant eyes. Black-yellow-orange concentric circles. Bull's-eye.

Then it was over, and they sauntered down the beach, a low chatter among themselves. I sat back on my heels and studied the waves. I couldn't say we communicated, or related; I couldn't say that any trust was gained. I could say that they didn't fly off or raise their voices or act agitated. That they approached me. That I felt, ever so briefly, a mutual awareness without fear.

Communicating

South Culross

It is a rainy afternoon in Picturesque Cove. I sit inside the small dark cabin, reading by candlelight from David Abram's Spell of the Sensuous about how our language creates a perceptual boundary that has separated us from the natural world. I stop to let this idea sink in, and look out the window to the brighter light outside. The dark wood frames dripping green salmonberry leaves and my son.

Jamie stands on the porch in his green rain suit, cawing to crows in his sweet six-year-old voice. He makes another sound, higher in pitch, sharper, and repeats it. Then he opens the heavy wood door, enters, and says, as he begins to poke at and blow on the fire, "Communicating with those seagulls for a second."

restoration

1997

EVENING LIGHT sifted golden through tall spruce and hemlock, illuminated cascading water. The white gauze of water poured toward me, tumbling over boulders, pulsating through a narrow rock passage. As the tide continued out, the waterfall steepened, dropping as much as ten feet, rushing to make the water level in the lagoon equal to that of Ewan Bay. When the tide turned and rose enough to begin filling the bay, the waterfall reversed, pushing water through the tight passage into the lagoon.

Standing in a skiff too large to fit through the passage, a bloom of orange lion's mane jellyfish floating beneath us, I remembered another trip to Ewan Bay. In our small inflatable, Andy, Jamie, and I had glided on the incoming tide from still water in the bay to the quick pull of waterfall to still water in the lagoon. We drifted in the oval lagoon, gazing up into dense forests concealed with thick-trunked trees draped in broad platforms of moss and long strands of old-man's beard. Rather than waiting four hours for the tide to turn so that we could pour back out into the bay, Andy worked the skiff up and over the rock ledge next to the waterfall while I carried Jamie through the woods.

Now, four years later, the forests of Ewan Bay and the other fingers of water along Dangerous Passage were newly protected from any large-scale development—especially clear-cutting. Chenega Corporation, the Alaska Native village corporation that owned this land, sold nearly sixty thousand acres of virgin temperate rain forest for $34 million, part of the $1 billion natural resource settlement managed by the Exxon Valdez Oil Spill Trustee Council. Most of the land would now become part of the Chugach National Forest.

I was here with a guide and a photographer to write this silver-lining story eight years after the spill. I'd met our guide, Roger, on that trip with Andy and Jamie. He and his family had lived here for eighteen years, running a small floating lodge in nearby Paddy Bay and serving as caretaker for the Chenega Corporation.

The sole purpose of the $1 billion fund was to try to restore the land and wildlife along the thirteen hundred miles of shoreline damaged by oil. It was not for compensating people harmed by the spill—a separate class-action suit addressed that. The money was for the place and its wild inhabitants. To restore the wild, reverse the damage.

This habitat acquisition was intended to protect the heart of the spill zone from darkening. All of the oil had flowed past here, and much had come ashore here—more than eleven million gallons of toxic crude in one of the most biologically productive areas of the Sound. Most of the Sound's tens of millions of salmon funneled through here in spring, leaving lives in the Pacific Ocean to spawn in the streams where they'd hatched; thousands of birds migrated through in spring, fanning out across Alaska to summer nesting grounds. The Sound's largest concentrations of Steller sea lions, orcas, and humpback whales congregated here as well, to feed on the abundance of krill and fish, to mate, to give birth.

Here, too, residents of Chenega Bay, a small Native community of mostly Aleut descent, had struggled to rebuild after a locally generated tsunami from the 1964 earthquake washed away their original village, taking with it twenty-three of their people. The Old Chenega village had been abandoned, the survivors scattering to Anchorage, Tatitlek, Cordova, Valdez. It took two decades for the community to relocate to a new site twenty miles away on Evans Island, two decades for the survivors to reunite. Then, twenty-five years after the earthquake, once more on Good Friday, oil covered their waters and their shores, threatening the existence of their village and their lives once more.

Roger took us past tsunami-swept Old Chenega on our way east to Sleepy Bay, a north-facing curve of beach on Latouche Island. In a protected cove, the abandoned village faced Whale Bay, Icy Bay, and the Sargent Icefield. Icebergs from Tiger and Chenega Glaciers poured toward it, and I recalled having flown over it, watching from above as three humpbacks glided through the ice-strewn waters. It was beautiful, a good place for a village—until the tsunami hit. Now it was abandoned, only the schoolhouse, high on a grassy hill, visible from a distance. Now the remaining fifty-six villagers gathered there once a year for three days to remember the dead. Not that remembering was hard.

Roger told me about the brother of a man whom I'd met earlier. "When he saw the wave coming, he grabbed his two kids, the boy by one hand, the girl by the other, and started running from the beach up the hill. He held tight, but they were yanked from his grip by the force of that water. They were never found. He took to drinking, died a couple of years ago, I think." Roger bowed his head. "It's sad, so many of them lost family, especially kids, because the kids were all playing down on the beach when it hit."

On this day, the waters were calm and glassy; as we passed Whale Bay, Roger told me that most of his clients were "fish killers." They came for one thing: to catch as many large fish as possible. He always took them to Chenega Glacier anyhow, though they complained all the way about missing time fishing.

"Once they're there, once they get to see the ice calve from the glacier, hear the sound of it, they forget all about their fishing poles," he said, "and they tell me, 'Thanks for making us do this.'"

Glaciers glistened in sunlight as we passed a pod of orcas feeding along Knight Island Passage. I could almost believe that all was well here now, that the effects of the earthquake, the effects of the oil spill, were but faint memories that cast only the lightest of shadows over this place.

But at Sleepy Bay, multicolored booms in three parallel half-circles cordoned off a section of beach. A dozen people, dressed in bright-yellow rain gear though there wasn't a cloud in the sky, milled around onshore, wielding buckets and long pipes. Though the scene contrasted sharply with all other beaches we had passed on our way here, it was also nightmarishly familiar. In 1989, nearly every beach looked like this: Wrong. Violated.

Onshore, we were immediately approached by three white men wearing hard hats and jeans. At first they seemed guarded, asking us who we were, what we wanted. But then they relaxed, posed for photographs, and showed us the latest oil-cleanup technology—new devices designed to flush the oil out from beneath the rocks, oil that months of hand scrubbing, summers of cold- and hot-water washing, years of biochemical treatment had not budged.

Sleepy Bay had endured all attempts to purge it of oil. In 1990, I had witnessed bulldozers moving rocks, digging into the beach to uncover oil for hot water to wash away. The beach wasn't cleaned so much as rebuilt, left scrubbed and sterile. And still it held oil. As did beaches on Knight, Green, Montague, and other islands in the southwestern Sound. Most beach cleanup had only been cosmetic, creating a dangerous illusion that the Sound was clean, even though less than 10 percent of the oil had been recovered. Now, though other wild-salmon stocks had finally returned to the Sound, the stream at Sleepy Bay was forsaken; it should have held hundreds of spawning salmon.

Just past the barren stream, ten residents of Chenega Bay, in uniform yellow suits and thick rubber gloves, were once again trying to clean this beach. They worked silently, not even looking up from their task, moving hoses, aiming water sprays, wiping off rocks. Before the oil spill, they and their families had come to this beach to collect mussels and kelp, to fish for salmon, to hunt deer. Now they bent to another task, their movements slow, as if they were dazed.

I walked over toward them, and one looked up and met my eyes. He introduced himself as Larry Evanoff, a name I recognized. He and his wife, Gail, had worked hard to relocate and reconstruct the community after the earthquake. Now he was working on another kind of restoration. He turned over a rock and showed me oil glistening beneath, as fresh as if it had washed up the day before.

He didn't want to answer my questions about the habitat acquisition. Not all of the villagers had supported the sale; it felt too much like another loss in a long line of losses. Still, village corporation executives, charged with earning money for their shareholders, had seen only two options: clear-cut or sell. Since the residents retained exclusive hunting and fishing rights on these lands, many believed the sale to be in their best interests, as well as in the land's best interest. It was an uneasy choice.

On our way to this beach, Roger had told us how oil spill workers had defiled the old village site: the plaque with the names of the dead chipped away, the old schoolhouse stricken with graffiti—OIL SPILL CLEANUP 1989—adding insult to injury. But still Chenega Bay residents were here, at Sleepy Bay, trying to clean the oil. Was it the money they made, now that their subsistence future was in jeopardy, or did they still believe this technology would give them an oil-free beach?

When we left, the villagers sat in a row on beach logs like cormorants on a rock ledge, eating lunch, staring out to sea.

I would recall this scene a year later, at a meeting of the Exxon Valdez Oil Spill Trustee Council in Anchorage, where discussion focused on what the Trustee Council ought to do to mark the tenth anniversary of the oil spill. A woman in the back of the room rose to speak; she was a resident of the Native village of Tatitlek.

"Can't we call it something other than an anniversary?" she asked. "That word means a celebration of some event, and there is nothing to celebrate here."

After Sleepy Bay, Roger set a course for Jackpot Bay. Off Dangerous Passage, Jackpot spreads into a series of waterways, some connected like a string of pearls by narrow passages, another widening to embrace a massive waterfall, and one curving back to a small stream. When I'd been there before, the only evidence of humans I'd seen had been two sportfishing boats.

This time, a small settlement had risen near the mouth of the stream: three large wall tents, a couple of buoys on the water, a gas can on the beach, a bright-blue tarp covering three fuel barrels, and trails crisscrossing the small headland. This was now a research camp, Roger told me, peopled all summer.

We boated ashore and were met by a young man in shorts pulled over long underwear, a T-shirt, and a baseball cap. A graduate student, Phil was here for the summer helping with the pigeon guillemot project. We followed him down a well-trodden trail to a wall tent, and stepped inside. He introduced me to two other researchers, and I asked about their projects, funded by the Oil Spill Trustee Council as restoration work.

They were studying chick production and growth of a colony of pigeon guillemots on a small mushroom-shaped islet in the middle of Jackpot Bay. These small black-and-white seabirds had been declining since before the oil spill, and it was estimated that up to 20 percent of the population had died from the oil spill. This colony was the largest left in southwestern Prince William Sound. Every morning, when the adult pigeon guillemots were out feeding on the water, these researchers climbed the island, found tunnels in which the birds nest, and reached their hands and arms into them, sometimes up to their shoulders, groping for eggs. They counted and marked them. Later they would capture and band the chicks that hatched.

I asked what they hoped to learn from the project, and how it might help pigeon guillemots recover from the oil spill; they weren't sure. I asked whether protecting the surrounding forests from clear-cutting would help pigeon guillemots, but again they didn't know. There was an uncomfortable silence, then the sound of another motorboat approaching. Phil said it was the project director.

"Oh, I'd like to talk with him," I said. Perhaps he would be able to tell me more.

One of them took the skiff out to meet the boat; I waited for nearly a half hour, but neither she nor the project director returned.

"He's probably giving her a hard time for talking with you," the other two told me.

In my bunk that night, I was so upset I couldn't sleep. I kept thinking about my earlier visit here in 1993. It had been a long-awaited trip to a place I'd wanted to visit since my first summer in Alaska. We had drifted among icebergs, hiked up mountainsides through meadows, splashed in clear-running streams, watched sea lions porpoise around our boat, seen a pod of orcas. It had been magical.

I had known I was there only four years after the oil spill; I was aware that there were fewer harbor seals, sea otters, birds. But I had believed the Sound was recovering. Wounds needed time, undisturbed time, to heal. I knew this all too well now, as I struggled with the wounds of my divorce. I thought that after the frenetic two-year, $2 billion cleanup and $173 million damage assessment, the Sound was finally getting that undisturbed time to restore.

I had been wrong. Every day I was out with Roger, we saw more people, boats, cleanup crews, and research camps than I had in 1993. Before the spill, only a few researchers worked in the Sound, tagging salmon or counting whales or tracking sea otters. Now, with funds from the Exxon settlement, a legion of scientists studied pigeon guillemots, river otters, sea otters, mussels, harlequins, salmon, herring. Now, instead of damage assessment, it was called restoration. The spill and the $1 billion settlement spawned a new industry whose center was here.

I wondered why the bird researchers had been so hesitant about answering my questions. Perhaps they weren't hiding anything; perhaps that was the problem. Perhaps there was no benefit, no restoration, for their research—just data and banded birds. I couldn't see how pigeon guillemots, or any other of the Sound's wildlife, were now more protected from oil spills; I couldn't see how their lives were better or safer.

Lying in bed, relentlessly awake, I ran through a list of the other restoration projects the Trustee Council had funded. Surf scoters shot and set adrift; harlequin ducks and sea otters implanted with transmitters. Harbor seals and Steller sea lions—now listed as endangered—were declining before the spill, likely from diminished food sources caused by overfishing. Oil-slicked haulouts and rookeries exacerbated matters. Shortly after the spill, nearly two dozen seals were "collected" so their stomach contents could be studied. Many others had since been captured and burdened with radios and antennas on their backs.

Restoration funds also went to the Seward Sealife Center, a new aquarium touted as a research facility where tourists could see puffins and river otters and sea lions. Of the $50 million it took to build the aquarium, most came from the Exxon settlement. Each year, millions of dollars from the fund would go to support it. For one research project at the center, healthy river otters were captured and caged and fed oiled food, after which their blood was tested—all this to give researchers a benchmark for interpreting the blood samples they'd already collected from otters in the Sound. In another project, healthy harlequin ducks from nonoiled areas were likewise captured and caged in the Seward Sealife Center; they were to be released to the wild after the research was complete, but they had contracted a virus and were euthanized. Scientists didn't know whether the virus was contracted while in captivity or whether they already had it at the time of capture.

The money was used to fund other buildings besides the aquarium; nearly every community in the spill zone had a new facility from these funds. Kodiak got funds for a multimillion-dollar industrial technology center, Seward a commercial shellfish hatchery.

The Trustee Council set aside $140 million for a "Restoration Reserve," money that could be stretched to last for decades to, as one Trustee Council member said, "fund research by scientists who are still in grade school." An esteemed former legislator suggested using the reserve for science education throughout the state. Letters from University of Alaska professors requested funding for a variety of projects, many of them making absolutely no mention of how these projects were linked to restoration in the wake of the oil spill.

Restoration was defined as restitution for a loss, as returning to a previous and more desirable state, as renewing, giving back. All this counting, tagging, radio-implanting, all these projects, research, buildings—I could not fathom how they would help restore the wild.

©

It would be another year before I learned that there was much more to lose sleep over than the failure of restoration. At a panel discussion about the *Exxon Valdez* oil spill at a 1998 environmental history conference, Walt Parker, who had worked in oil transportation safety issues since the inception of the Trans-Alaska

Pipeline in 1977 and had chaired the State of Alaska's Oil Spill Commission, warned that the risk of an oil spill in Prince William Sound might be greater now than in 1989.

Oil spill prevention and response systems in the Sound had become among the best in the world. Escort vessels accompanied laden tankers from Valdez to Hinchinbrook Entrance, upgraded U.S. Coast Guard radar enhanced vessel traffic monitoring, speed limits were lowered, communications protocols had improved, and a full cadre of response equipment was at the ready. Most important, the Prince William Sound Regional Citizen's Advisory Council, a watchdog of both industry and government, had been formed.

But the weakest link in the system remained the tanker fleet itself—aging, still mostly single-hulled, increasingly suffering structural failures. The Oil Pollution Act of 1990 had mandated a phaseout of single-hulled tankers by 2015, but the industry was being maddeningly slow to respond. What's more, the pipeline itself was older and less sturdy.

"It was an old fleet in 1989," Parker said, "and now these aging ships have sustained another decade of battering from some of the roughest waters in the world."

At the same panel discussion, the science coordinator for the Trustee Council said, "Most of the recovery has and will come from natural processes." The research, he said, "provides information that will enable us to sustain the ecosystem over time."

I was stunned. The best we could do for Prince William Sound was to prevent such a spill from recurring, but we weren't even doing that. Instead, we were flushing beaches, sticking our hands in burrows, rounding up and capturing wildlife, and implanting transmitters. I couldn't help but wonder if all this science merely gave us the illusion that observing animals somehow helped them recover. The bald eagle population had recovered from the spill without any help from science—all science did was record it. Few restorative management decisions had resulted from research. River otter trapping, for example, hadn't been restricted in the southwestern Sound until after the river otter was moved to the "recovered" list.

Rick had become increasingly outraged at the amount of scientific research and monitoring funded from the settlement. This kind of basic research was what state and federal agencies should be doing as part of the routine fish and wildlife management, he said, not as restoration from the oil spill. He decided to file a motion in federal court to require an independent review of the Trustee Council's spending so far. We spent the better part of a weekend—Rick, his friend David, and I—typing up this "motion to intervene and compel compliance" in the proper legal format, in an attempt to steer the process back toward restoration. The court denied the motion.

If the Sound was to have a chance at recovery, I now believed, the best restoration would be to simply let it be. Limit the number of people in the Sound, the number of boats and camps—a permit system, like in Denali National Park. I'd be willing, even if it meant I couldn't go there every summer. In the name of restoration, I'd be willing.

But things were heading in the opposite direction. The state was building a road between Portage and Whittier, a road that would share one tunnel with the train, putting Prince William Sound within a ninety-minute drive of Anchorage. This was expected to increase visitation to the Sound to nearly fifteen times what it had been in 1989. There would be more boats than river otters, more people than pigeon guillemots. Only four lines in the road's Environmental Impact Statement were devoted to effects on the Sound, though even the bird researchers in Jackpot Bay, reticent as they were to talk about their own work, said the road might do more damage than the oil. Despite their charge to restore the Sound, the Trustee Council did nothing to prevent this next disaster.

It was hard to accept limits, limits to what we could do, to what we thought we could do. It was hard to accept that the best thing might be to do nothing.

๑

On our last morning, Roger took me to a small beach on Chenega Island. I walked into the forest, following a small stream crisscrossed with fallen logs and bending branches. Abandoning the stream, I followed an animal trail around two large boulders and up a steep bank. Sounds were muffled by thick moss: at my feet, on the branches, on the trunks around me. Every limb I touched felt mossy soft, wet and green. Walking was slow, for moss hid a tangle of fallen limbs and rocks.

I looked up at shafts of light pouring down upon small patches of the forest. So little light and sound in the middle of the day. One tree, larger than the rest, held several large moss platforms in its arms. I wondered if any were the nests of marbled murrelets. Small brown seabirds listed as threatened in much of the Pacific Northwest, their population in the Sound had dropped by 70 percent in the past twenty-five years, but they were still one of the most abundant seabirds.

Deep in the darkness of such stands, marbled murrelets nested. They were small, inconspicuous as they bobbed on the ocean. They spent days out at sea, then flew back into the old-growth spruce and hemlock forests, reaching speeds of one hundred miles an hour, little bolts of brown feathers darting among thick stands of rain-forest trees. Zip. Into the forest. Zip. Out to sea, to feed on the fish. Like needle and thread, sewing together land and sea.

I paused at the base of the tree, and, not able to find steady footing, grabbed hold of the trunk. I leaned into it, tried to encircle it with my arms, but could not. It was more than six arm's lengths in circumference. More than five hundred years old, this tree was now protected. An earthquake might fell it, a tsunami, but not a chain saw. Even if a marbled murrelet did not now nest in it, one could. Even if a river otter didn't make the trail I followed back down to the stream, one could.

It was the possibility of the wild that gave me hope. That this purchase, and others like it, left open the possibility for the wild to live unhindered. That sounded like restoration.

@

Habitat protection was the only thing the Trustee Council had funded that could help the place without inflicting more damage. Like double hulls and better-trained crews, it constituted prevention. But as a friend once said, if we considered the Sound a patient, then we ought to remember the healer's Hippocratic oath: First, do no more harm.

The Sound's forests began to fall to the chain saw at the same time as the oil flooded its waters. Preventing more harm was the impetus for the quick settlement. It still took years—years of negotiations, public testimony, meetings—to protect these forests. Throughout the difficult negotiations with private landowners and the Exxon Valdez Oil Spill Trustee Council, the question Rick kept asking was the standard against which every restoration action should be measured: "What's in the best interest of the ecosystem?"

This was the question I continued to hold before me, no matter how surreal and discouraging the process. At one Trustee Council meeting, I heard myself making this same plea again—do no more harm—but this time, having learned much since my first encounter, I chose my words purposefully, confident in the power of language to burn through logic-based systems. I wanted to create indelible images, knowing full well that they would most likely be met once more with silence.

I spoke into a small beige speakerphone in the middle of an empty conference table, in a windowless room on the top floor of a downtown Anchorage office building. Around me was a handful of others, all of us here to advocate once more for habitat protection. Most of the Trustee Council's six members listened in from their offices in Juneau, Anchorage, and Washington, D.C.

"This is not a logical appeal. I couldn't sleep last night. I kept hearing a tree falling. I kept hearing the cracking thunder of a great trunk breaking and then the thud of it hitting the forest floor. I felt the same sense of doom I did nearly six years ago. I couldn't sleep then, either. I kept seeing a sea otter coated in oil, shivering, and then finally sinking like a stone to the ocean floor. What can I say? Forgive us, Lord, for we know not what we do? No, we know what we do, and we know that there is no excuse for cutting one more tree in Prince William Sound."

With habitat protection, coastal forests slated for clear-cutting could be saved. Forests were integral to recovery and inextricably linked to oiled beaches and waterways. Forests harbored streams where salmon eggs hatched, the fry returning to the sea. Forests harbored birds that fed upon the fish in the Sound. Forests sheltered fragile intertidal areas, nurseries where freshwater and saltwater met. Forests were entwined with the sea in mutually dependent relation.

Restoration funds eventually protected seven hundred thousand acres in the spill region, completing millions of acres of intact ecosystems. The forests of southwestern Prince William Sound were protected before they were cut, but not all of the Sound's forests had been so lucky. Negotiations with two other landowners had failed to prevent thirty thousand acres of clear-cutting. If the focus had stayed on true restoration, on the oath of first doing no more harm, rather than on science, these forests on the eastern side of the Sound might have been saved.

Habitat protection ultimately restored what we most needed to restore: our relationship with the place. Not through management; not through abstinence. Restoring our right relationship with Prince William Sound required learning, or remembering, a way to be in the natural world that didn't desecrate or overrun, but that maintained and respected.

This is the way of attentive love: action based on the love that comes from being aware of what the place and its inhabitants need and desire. Attentive love requires an ethic of humility; it requires faith in ourselves and in the beloved. Rejecting excessive control, it reveres the process of life. At times the way of attentive love is inaction: sometimes it's best to do nothing. Examples abound: the genetic scientist Barbara McClintock said she listened to the corn; biologist Jane Goodall let herself be guided by the chimpanzees. My guide, Roger, showed it as well.

I recalled our first day with Roger. Along Dangerous Passage, we had stopped near a small island in Paddy Bay.

"There," he pointed to a tall Sitka spruce, "in the top. See the eagle's nest? This is the thirteenth summer they've nested here."

At the mouth of Eshamy, he pointed out a spit of land. "A few falls ago," he told us, "I caught five bear hunters camped there. I chased them off, told them no hunting allowed on this land." He told me stories about catching kayakers littering, a sailboater dumping used oil, hunters after bear and sheep.

One evening, we walked up to Eshamy Lake. Standing at the edge, staring into clear water, he told me about walking over the logjam at the lake's outlet.

"I had the feeling I was being watched," he said. He stood still for a moment, then looked down and saw two pairs of eyes, two river otters, looking back.

On that last morning, after leaving the forest where marbled murrelets might nest, Roger radioed a scientist I'd been hoping to meet with, one who was studying river otters. She was on Knight Island, capturing river otters in live traps, then taking blood and comparing samples from oiled and nonoiled populations. I asked her the same question I'd asked the others: would protection of these forests help river otters?

"Yes," she said. "Absolutely. Protecting forests from logging is the best thing we can do for river otters."

I couldn't see her, could barely hear her voice, but in her words was a genuine concern for the river otter's well-being, an openness about the importance of protecting the forest. No wariness, competitiveness, no sense that money used for habitat protection took away money used for science.

I couldn't comprehend doing her work, and didn't know how it might help river otters. But in her voice, her words, was a compassion for the animals she studied, an attentive love for river otters.

I tried to imagine the Trustee Council meeting to decide about research funding not in a conference room in Anchorage, but on the beach at Sleepy Bay, or by the stream at Jackpot Bay. They had never gone to the Sound as the council. When Rick, as marine adviser for Prince William Sound, extended an invitation for a site visit, they responded that they were too busy. What might have happened, though, if they had each spent time watching harlequin ducks in their habitat before deciding whether to fund another harlequin roundup?

What if we gathered each year as the earthquake survivors at Old Chenega did, remembering and honoring? What if, instead of holding a technology conference in Valdez or a science conference in Anchorage, we gathered out here on the anniversary of the spill?

If we listened, the Sound might tell us that we didn't need to capture harlequins, we didn't need to band pigeon guillemots, we didn't need to excavate beaches. The Sound might show us other paths of restoration.

As the plane took off and my view of the newly protected lands expanded and then grew smaller, I thought about some friends who had spent years observing sea otters in the Sound. "How would saving the forest help the sea otters recover from the oil spill?" I had asked them.

In winter, they had told me, when all the researchers had broken camp and gone home, when the cleanup equipment was stored in a warehouse, when the villagers were in their homes in Chenega Bay and Tatitlek, eating dried salmon, the reversing waterfall at Ewan Bay was rife with life. Lined with a bed of mussels more than a foot deep, the lagoon at low tide drew in sea otters. Just-weaned pups feasted on mussels in shallow, still waters, while their mothers hauled out and rested on the boulders by the waterfall. River otters, ermine, and other forest creatures sometimes came down from the forest. It was an oasis for the sea otters, especially the pups, from wild winter storms. It was a place of commingling, forest and sea life entwined. In the longest season, winter, the place was left to itself, there in the still lagoon, by the waterfall, which continued to flow back and forth with every tide.

Clouds

Port Fidalgo

Out the window of this Cessna 206, the weekly mail plane to Chenega Bay and Tatitlek, I see clouds in the water, light and dark. The light clouds, blurring into gray-blue water, are blooms of jellyfish. Great blossomings of pale-white moon jellies, pulsations of light. The dark clouds, they are Pacific herring. Eight inches long, they swim in schools of a million or more, a sudden flash of their silver undersides confusing predators. In April their spawning turns the waters of bays and lagoons milky white; from sea and land and air come those who feed on their roe. More than forty different species—bald eagles, brown bears, humpback whales, tufted puffins—all depending on such small fish with such big lives. From above they are dark water; from below they are sky. I sit in the floatplane and imagine the view from down there, from beneath those millions of herring. One dark cloud flashes silver, a bolt of lightning in a sea of blue.

in this light

1997

IT WAS BIGGER, brighter. Its tail streaked farther across the night sky. Looking at it through binoculars, I had to move them slowly, steadily, panning like a photographer at a race, to see it all, to see the sunburst head and the tail that arched and fanned like a single wide brush stroke. This comet seemed so different from where I now stood than from my porch at home, but it was the same comet, Hale Bopp, that I had come to expect on clear nights, come to expect as much as I expected to see Orion's Belt and the Big Dipper, even though in a few weeks it would disappear, not to return in my lifetime.

To the west was the quarter moon, its dark half filled out with Earth shine and its sunlit half so brilliant that through binoculars its craters were visible. To the east glowed Mars, a red beacon in blackness. All this shone above me from the porch of a cabin on Green Island along the southern edge of Prince William Sound, in the middle of March. In just one week, in the wake of the spring equinox, this comet would reach its zenith, this celestial gathering would reach its brightest night, and this place would have eight years between it and the *Exxon Valdez* oil spill that coated the beaches before me and killed throngs of seabirds and sea otters and fish in the waters around me.

Green Island. I had never been here before, but I'd heard of it. In April 1989, a month after the tanker hit Bligh Reef, birds started arriving by the thousands, some to stay and some just passing through on their spring migrations to northern nesting grounds. People awaited them, firing gunshots into the air to keep the tired birds from landing, for instead of the respite the birds sought they would find only deadly oil.

Now a comet spiraled overhead within a few million miles of Earth, Mars rose in direct opposition with the sun, and the moon headed toward a near-total eclipse. Looking up into the night sky, I wondered: what did this mean, this brilliant celestial event and this dark anniversary converging?

I hadn't planned this trip to coincide with these prodigious events, but the oil spill had profoundly affected me, as it had irrevocably harmed the Sound. Memories of those dark years haunted me; I couldn't separate place from event. At times the entanglement tightened its grip into obsession.

Underlying this despair, hard as bedrock, were love and desire. Prince William Sound was more dear to me than anywhere else on earth, and I craved time in it. I had wanted to come to Green Island for years.

Jamie and Rick and I were staying in a cabin on a bight of land on the western edge of the island, sandwiched between two beaches. The front beach faced north toward Montague Strait, with Naked Island in the distance; the back beach faced a wide, quiet finger of water flanked by the main body of the island and a narrow spit. Green Island lacked the towering ice-covered peaks of Knight and Montague Islands to either side of it; instead, true to its name, it remained green year-round with a dense forest of spruce and hemlock. It was low lying, marshy, dotted with ponds, perfect for migrating birds.

Morning. On the beach facing Montague Strait, waves crashed against long rows of tangled kelp and sea grass and driftwood that had been washed up in winter's storms. Wind roared through the trees, their needled branches lashing at the cabin windows. The seas, though, laid low enough for us to venture out in our kayak. We pushed off into the wind, hoping to paddle around the point to see Montague Island. It and Green and Knight Islands to the north had caught the bulk of oil in the arms of their fjords.

I was anxious to lay eyes on Montague, having seen it only from the air. Choppy seas made for slow going, though, and while March heralded spring equinox, here it was still winter: snow crusted the shoreline, ice covered freshwater, and the slightest breeze sent a chill. We were almost to the point, so close I could imagine rounding that steep headland and coming into full view of that magnificent island of snowcapped peaks, forty miles long and abounding in brown bears and Sitka black-tailed deer. But Rick worried about the building swells, and Jamie whimpered with cold. We turned our backs to Montague and headed to the cabin. Struggling with disappointment, I helped Rick beach the boat and followed Jamie down the beach to the headland nearest the cabin.

Low tide, every crevice, every cirque in the rock, held saltwater filled with life. Jamie and I climbed from one tidepool to the next, peering in. Once our eyes adjusted, we found rockfish, anemones, sea stars, sea urchins, barnacles, limpets, and snails among forests of sea lettuce, kelp, and eelgrass. One long, narrow pool was packed with sea stars, orange and burgundy and yellow stars among bright-green waving eelgrass. With a long yellow float that Jamie found, I pried up one red sea star, flipping it over so that its white belly showed.

"Watch and it will turn itself back over," I whispered. A few pale tentacles waved around, reaching out for a firm footing. Gradually more and more tentacles reached, rippling across the thick arm like swells in the ocean. Almost imperceptibly the legs stretched and bent, and the body of the sea star metamorphosed onto its side.

"C'mon, let's go," said Jamie after a few minutes. "I want to climb up to there." He pointed to a rock face that led straight up to the forest edge.

"We have to wait. The sea star is vulnerable this way. I don't want anything to happen to it because we turned it over," I told him.

"What could happen?" he said, interested again.

We continued to stare into the tidepool as tentacles reached and curled. The change was slow but certain. It was almost, except for one white tip, right-side up again, squeezing itself in among the other stars, when we started climbing the rock.

Hand pressed against smooth gray rock; fingers reached for a small hold where rock jutted out, where rain eroded a dimple, a crack; eyes focused on patterns of granite inches away.

"Put your foot here, and then put your hand here," my five-year-old son told me. He led, watching me closely as I followed him up the boulder. "Go the same way I do," he said.

I listened: he found the best route.

We crouched on top of the boulder, feet on soft moss, shoulders touching, and looked out from the headland to the waters of Prince William Sound. I put my arm around him, stroked his tousled, sun-streaked hair. The waves of morning had flattened to ripples on the surface. The rock we'd climbed a few minutes before to look into tidepools was slowly surrounded by the incoming tide. I pointed to it.

"Look," I said. "See how that rock is becoming an island? See how the lay of the land changes every moment?"

Picking our way across the rocks not yet submerged by the tide, we returned to the sandy beach. Jamie found a wave-washed spruce limb and prodded the kelp berm at the high-tide line, digging for treasure among the debris of storms. Dozens of tiny beach hoppers sprang in all directions as he pushed the tangle of seaweed and shells. He crouched and cupped his hands, hoping to catch one. I located a boulder half-submerged in sand, smooth, gently sloping, facing the sun. Leaning against it, I closed my eyes as heat arose from black sand and rock. My head and back were cradled in the curve of rock; my feet, free of boots and socks, dug into warm sand.

"Mom, come look at this," Jamie said. I got up, too quickly, and saw stars, lost for a moment in a universe of my own.

Jamie pointed to a dead sun star. Its bright-orange color was fading, bleaching away, and its skin peeled off in strips, eaten by the beach hoppers and beetles hidden in the kelp berm. We turned it over to expose the skeleton beneath, white cartilage radiating out in twelve thin lines, each with matching patterns of thinner strips across, the whole thing like a snowflake magnified. By the next day, there would be no skin, no color, no body. Just white arms radiating upon black sand, a star skeleton.

۞

Afternoon. Spruce shadows lengthened over the beach, so we walked over to the sunlit lagoon on the other side of the cabin. Jamie played at the water's edge, throwing in rocks and singing. I started to read, but the words on the page couldn't hold me, so I joined Jamie at the water's edge.

Something moved there, something small and translucent. A feather? An embryonic stage of a marine animal? It wiggled back and forth, propelling itself slowly through still water. Its body was soft and oval-shaped, with flaps on either side and a row of three finlike appendages. It was ethereal. I showed it to Rick and Jamie.

"I think it's a nudibranch, a sea slug, called *melibe,*" Rick said.

"I thought about that," I said, "but I've only seen nudibranchs in photographs of tropical waters. I didn't know they lived this far north."

Suddenly I remembered seeing something in the water when we were paddling in this lagoon the day before. The image returned as quickly as a flash of light. We were hurrying to get back before dark, and I hadn't stopped to look longer or mention it to anyone.

The water was so clear; I had never seen it so clear, not in all the times I'd been in the Sound. I could gaze deep into it, as if I were looking through the lens of a telescope, no surface movement to distort the image; the water seemed, in fact, to magnify what moved within it. Long, thick stands of green kelp waved up from the bottom, dark-green palms arising from burgundy stalks. Attached to and floating among them drifted tiny translucent pink things, hundreds of them, like underwater flowers blowing in the currents instead of the breeze, blossoms that had blown off their branches to float through forests of kelp.

Clouds rolled in again that night, obscuring Hale Bopp and Mars and the waxing moon, but they had dissipated by morning. We paddled out from the front beach toward the western point, hoping to reach some islets on the other side.

I dipped my oar into flat, clear water, looking fifty feet or more into it, once again astounded by clarity. Colors intensified where the sun's rays filtered far into the water. Jamie sat in the middle again, calling out excitedly at every discovery.

"Look, another sea star! A purple one!" he cried. "And there! An orange anemone!"

All was still and quiet, dreamlike, except for the finest sounds of forest and sea. As we paddled, a pair of magpies, their black feathers an iridescent blue in sunlight, followed us from the shore, hopping from rock to rock, flying a few feet, dropping onto a new perch, measuring our progress.

"Our escorts," Rick said, and we laughed.

At the point, we paddled through some sea stacks, coming right up on a couple of sea otters and a raft of seabirds. Among them were cormorants and guillemots, and some ducklike birds with striking silver and white and burgundy markings. They were harlequin ducks, birds I had never before seen, but had heard about.

Harlequins had suffered greatly from the oil spill, and they suffered still; they had yet to breed successfully, and females were dying at higher rates than usual in winter. That confusing mixture of sadness and pleasure found me again: sadness over the loss and their continued suffering, pleasure at seeing them here nonetheless, seeing their brilliance.

I wanted to reconcile these feelings, but how? Not by glossing over pain, not by denying joy. How, then? Like oil and water, they didn't mix.

At the headland, we pulled the kayak up on some boulders covered with popweed and walked up to the high-tide beach. Strong, thick blades of beach ryegrass were just starting to push new leaves through the layer of snow and last year's grass. I sat in the sun, basking like a seal on a rock, until I was startled by a sound at once strange, familiar, and unfortunate.

An outboard engine. A small aluminum skiff carrying three people in bright-orange float coats zoomed toward us. The person in front held a clipboard; they were scientists of aftermath. Perhaps they were counting the harlequins, or maybe sea otters or scoters.

The roar of their engine amplified the roar of my sorrow and frustration and rage over what had happened and continued to happen here on Green Island. Every day others like them were out counting and capturing, prodding the lives of these animals.

They motored through slowly, staying far out from land; the otters dove, and seabirds flew off at their passage; they gunned the engine and quickly became a small point on the horizon.

When the wake dissipated, we climbed back in the kayak, paddling in the same direction as their passing. The birds and otters didn't dive or fly off at our appearance, but I couldn't help but imagine how the clamor of noise and wind from the floatplane that had brought us here must have disturbed them.

The islets came into view, a dozen turrets of land tufted with wind-sculpted spruce and skirted with pebble beaches. We stopped to eat and skip stones. A few harbor seals swam nearby, alternately spy-hopping, gliding, and diving, as if to play hide-and-seek, though they did all the hiding. I wondered what it would be like to see one of the harbor seals or sea otters underwater, to glimpse their rounded bodies, so awkward on land, slipping gracefully through the sea. The clarity of the water, the sun in the sky, the wild lives all around—it almost seemed possible. But I didn't voice my desire; it seemed too much to ask, a thing only of imagination.

All day long we paddled in the equinox sun. We talked quietly, briefly, our voices as languid as the water's surface. Puffins ahead of us flapped out of the water, then landed again behind us, settling easily. Every now and then, a salmon leaped through the glassy surface. Life vibrated around us, between us, within us.

Late in the day, beyond the islets and the seals, in the middle of Montague Strait: a sudden spume of water and a flash of black fins backlit by the sea. A pod of orca whales, eight or so, arced its way northward. A large male porpoised out front; a group of females and calves moved close behind. Their white saddle patches glowed pink in the sunset. They surfaced again and again, heading toward the snow-capped Chugach Mountains in the distance.

We rounded the last headland and came back into view of the

quiet lagoon leading to the cabin. To the left, a sea otter perched on a rock. She stretched up, glanced our way, then slipped into the water with a small splash. I considered paddling over quickly in hopes of seeing her underwater—but this day had already given so much. I watched for where she might resurface, hoping at least to catch a glimpse of her again.

@

Now. Look. The sea otter under the water right beside me, just inches below the surface. Sliding around the front of the kayak, bubbles streaming from her fur and from her nostrils as she streaks through water. Air escapes from the densest fur in the world, pouring out in long spirals of bubbles. She moves beside the kayak like a comet in the night sky. She zigzags back, rolls, and stops in midroll, curled into a half-moon. Sea otter looks up through the water at me. Then she rolls again, shimmers away under the water, and is gone. And I, I am saying, over and over, "Look! Look! Look!"

@

Morning. We walked the front beach, only a few hours before the floatplane would return for us. The sun-star skeleton was nowhere to be found, washed away in the night's high tide. I imagined its fragile bones scattered along the shore.

Past the smooth rock I leaned against, Rick's tall frame was crouched over rocks, sun glinting off his fair hair. Head bent, he was turning over rocks, finding what he was after: oil. He showed me splatters on boulders that looked like asphalt gone awry. He pulled back one big flat rock to find a puddle whose surface rainbowed when exposed to light.

"I've got to bring some of this back," he said. "People all over the U.S. need to see this. They need to see that Exxon's lying about how clean the Sound is."

He collected some in a black plastic trash bag, wiping his hands, those long, graceful fingers, on his coat as he worked. Those dark streaks of oil on his green coat, they would be there forever. I knew. We both had pants stained with crude. I didn't want to see oil-stained clothes again. I ran to the cabin and returned with a towel for him to use, too late.

The questions I had asked our first night here resurfaced, looking up at me through that reflecting pool of oil. What does this mean, all this beauty and horror mixed together?

Look: moon, water, comet, trees, sea star, otter, light.

Otter

Mink Island

We beach the boats and find a narrow trail, follow it through devil's club and blueberry, up. Out onto a muskeg meadow, to a faint trail of pale-green sphagnum moss smoothed over. Not for footprints but as a slide, a trail leads from one small muskeg pond to another, some no bigger than a foot across. The ponds are still water, deeper than they look, sinking soft brown bottom. We take the faint slide trail up through one then another meadow, to a ridge lined with blueberry bushes and dwarfed mountain hemlock. On our bellies, we peer over and see, in the next meadow, in a bigger pond, three river otters splashing and diving. Three shining brown bodies, slender as my arm. The biggest has a salmon in its mouth. It dives in a sinuous arc, a soft splash wafting across the meadow to us. In seconds, it pops up empty-mouthed. The smallest slips under the water and resurfaces with the fish in its mouth. Then all three leap into the pond and are gone. I turn back to the way we all came. That trail we traced here, it was made by this otter family, a slide they'll follow on their bellies back down to the sea.

ab pod

1997

For five days I had been on board the *Lucky Star,* and for five days we had not seen the AB pod. We traveled up Montague Strait, through Knight Island Passage, and down Latouche Passage. We drifted and dropped the hydrophone; we heard nothing but distant engine noise. We saw humpback whales, Steller sea lions, Dall's porpoises, harbor porpoises, harbor seals, and sea otters, but no orcas.

"This is the longest dry stretch this summer," Craig said apologetically. "Ten years ago, we were on whales every day here."

I was in southwestern Prince William Sound looking for orcas, also called killer whales, with three marine biologists: Craig Matkin, Eva Saulitis, and Mercedes Guiterrez. Mercedes, a graduate student, was here for the summer from her native Mexico, hoping to take what she learned back to Baja California and the orcas that frequented those waters. For Craig and Eva, this was their living: studying the orcas of southcentral Alaska.

Craig had been doing this for nearly three decades. Up until a few years ago, the vast majority of his work took place in Prince William Sound, and focused solely on photo identification: taking pictures of the whales' dorsal fins to identify individuals and family groups, or pods, and through this to learn about behaviors and changes in the population.

For nearly five years, I had been trying to join one of these research trips. I first talked with Craig about the whales in 1988, writing a story on the AB pod's new and problematic habit of taking black cod and halibut from longlines as commercial fishermen hauled them up. That was also the first time I talked with Rick, interviewing him about a project he and Craig had worked on to resolve this problem. Since, I had felt an affinity for this family of whales.

Orca populations contain two genetically distinct groups: transients, who eat marine mammals; and residents, who eat only fish. It was the transients who had earned the whales the title "wolves of the sea." The AB pod, like most of the nearly three hundred killer whales in southcentral Alaska, were residents. Although transients and residents look the same to the untrained eye, they don't interact. Residents form large matrilineal pods and usually have a smaller and more distinct range than the quieter, less social transients.

From the start, I was attracted to the AB pod, resident whales that showed such savvy and friendliness. This was the only southcentral pod that had learned, and had persistently continued, to take bottom fish from longlines; this had also been the largest and most gregarious pod in the Sound. They were congenial not only with other orca pods, but also with humans—they were the whales most often seen by commercial fishing boats and tour boats alike. Craig and Eva had spent so much time observing them that they'd learned the individual traits of some, and given many of them names other than the research standard: a combination of letters and numbers (*A* for Alaska, *B* for the pod, and then a number to identify individuals). Some were named after a place in the Sound—Montague, Eshamy, Chenega. Others were named for some character trait—Sphinx, Splash, Blue.

I kept up with the AB pod, hearing about them through Craig and Eva's work. I'd seen orcas several times in the Sound, but usually no more than a few. What I'd seen were most likely transients from the AT group, or perhaps one of the smaller resident pods. I'd been wanting to see this unique family of orcas for a long time.

So far, however, I had seen only their photographs in the field copy of *A Catalogue of Prince William Sound Killer Whales.* Published in 1992 and organized by pod, this book contained photos of the dorsal fins of every known killer whale in southcentral Alaska. Researchers had identified these individual whales by their dorsal fins, by the varying shapes of white saddle patches and fins, by distinctive scratches and notches.

This copy was full of pencil marks that updated it every year. Some pointed to new notches in the dorsal fin; some marked the sex of the whale. There were new boxes drawn in that showed new off-spring. The most striking marks on the four pages of photos of the AB pod members, though, were the large *X*s struck through so many, *X*s that meant the whales were dead. *AB29: Missing 91. AB41: Missing 94. AB3: Dead 97.* The AB pod, once the largest and fastest growing of any pod in southcentral Alaska, had contained thirty-five whales in 1984; now, only sixteen of those thirty-five, plus six of nine calves born since 1992, remained.

Such mortality for a pod is unprecedented in killer whale biology. The typical pod mortality rate worldwide is 2 percent—or one whale every few years. Orcas have a life span remarkably like ours. They reach maturity between twelve and fifteen years of age; from then until around forty years of age, females give birth to a single calf every few years. Males can live into their fifties, and females into their eighties.

The AB pod's astonishing losses were mostly attributed to two consecutive sources: bullet wounds from longline fishermen, and oil from the *Exxon Valdez* oil spill. This most convivial of any pod was now the only one of the nine resident pods that had, over the past decade, declined.

©

I first met up with Craig, Eva, and Mercedes in Seward, the port town at the head of Resurrection Bay, just west of Prince William Sound along the Gulf of Alaska. They had been whale-watching out of Seward, but with my arrival we headed back to Prince William Sound where, I assumed, we would see as many whales as they had been seeing in Resurrection Bay. The only thing I was curious about was what the whales might be doing when we found them.

Two days before I arrived, Craig told me they'd had an extraordinary sighting of the AN pod. They were hiking a ridge on Fox Island when Lars, Craig's seven-year-old son, said he saw some whales coming into the bay.

"Well, we didn't believe him at first," Craig said, "but then we saw them. They were just charging into the bay. We ran down the hillside to the beach, and they came up and rubbed on the gravel, right at our feet."

"Wow," I said, "wish I'd been there."

"We'll see lots of whales, don't worry."

After three days out, I not only wondered if I would see the AB pod, but I also wondered if I would see any orcas the entire week.

At first, it was enough just to be in Prince William Sound, to explore some of the places along the outer edge that I had heard of but not yet seen. Places like The Needle, a rock spire jutting one hundred feet out of the water in the middle of Montague Strait, more than five miles from any island, like a fata morgana. We motored around The Needle several times, looking for whales. White guano and the perilous nests of a kittiwake colony blanketed the highest rock promontories; Steller sea lions, their brown skin blending into the wet rock, lounged on rock platforms nearer the water's edge. All around in the water were seabirds—kittiwakes, puffins, and pigeon guillemots. The sea lions, at least fifty of them, eyed us halfheartedly. A couple of bulls bellowed, a few waddled off to the other side of The Needle, and one slid-splashed into the water. But we found no transient killer whales trolling the waters around it, hunting.

Sea lions and other marine mammals know the difference between resident pods like AB, who eat only fish, and transient pods who eat them. When residents swim by, sea lions hassle them, Dall 's porpoise play in their wake, seals don't budge from their haulouts. One year, a Dall's porpoise swam with the AB pod all summer. But when transients are nearby, sea lions haul out and huddle high up on the rocks, trying to stay out of reach of powerful jumps and sharp teeth.

I had wanted to spend the week in Prince William Sound watching the AB pod. I had imagined orcas surfacing by the boat, so close

I could almost touch them, orcas surrounding us, as if the boat itself was a member of the pod. I spent hours on deck, staring at water, leaning out, standing so long my legs ached. But they were not to be found.

I had to choose between staying in the Sound and looking for the AB pod, or going to Resurrection Bay and finding other pods. It was a tough choice: the Sound or the whales; it was a choice I didn't think I'd have to make. But I voted for a return to Resurrection Bay, as did Craig and Mercedes. Eva was much more reticent, the lone dissenting voice, and I soon learned why: Resurrection Bay was crowded and noisy.

When we were a few miles out from Resurrection Bay, Craig heard from a tour boat captain who had seen a pod the day before—about fifteen whales, the AJ pod, he said over the marine radio.

Everything shifted. I put my book away and looked out at the endless expanse of the Gulf of Alaska bordered by the jagged mountains of the Kenai Peninsula. Craig's mood switched from irritated and indecisive to excited and happy.

"He's so much happier when we're on whales," Eva told me.

The last of the morning clouds lifted and left us in sunshine. As we rounded the last headland before Resurrection Bay, the water changed from flat calm to choppy—a wind from the head of the bay. Boats were everywhere, dozens of small sportfishing motorboats, sailboats, yachts, and large tour boats.

Craig got continual updates on the radio, and we were almost to the Seward harbor when we saw them: four tight groups of rounded black backs with white patches at the base of tall, curved fins. Their black skin gleamed in the sunlight, carving the waves. The backs of each group were in rows, dorsal fin to dorsal fin, arcing so slowly and with such synchronicity that the group seems to be porpoising in place.

"Group resting," Eva said it was called, a ritualized pod activity as close to sleep as whales got.

Resident killer whale pods comprised several maternal groups—mothers and their offspring—and could range from five to forty whales. This pod, the AJ, had thirty-two members. It was hard for me to count them, but there were at least twenty orcas. I gripped the rails, mesmerized, my heart pounding, staring at the whales, unaware of what was happening beyond them. Craig and Eva weren't.

Several small boats were around the whales, watching just as we were. Then one sailboat, running on engine, plowed right among them, almost on top of them. A once-tight group of resting orcas dispersed. On the radio, we heard one tour boat captain remark on the sailboat's behavior, adding, "They wouldn't do that if they knew who was watching them."

They were referring to us. The *Lucky Star* was well known, not only for studying killer whales but also for educating boaters on viewing whales.

Federal laws prohibited boaters from disturbing marine mammals in ways that altered their resting, feeding, and movement patterns, but enforcement was limited. Because they were out here every summer, Craig and Eva felt a responsibility to act if they saw boaters harassing whales.

"Long-term monitoring projects like ours are in essence a kind of enforcement," Craig told me as we motored over to the sailboat. Craig pleasantly and carefully explained to them the respectful—and legal—way to watch the whales.

He had worked with the tour boat captains for several years, convincing them to decipher which way the pod was traveling, position their boats far ahead of the pod, cut the engines, and wait for the whales to come to them. This worked. This way the whales weren't being chased; this way they chose whether to approach the boats.

Most captains followed Craig's advice. Too much disturbance could chase off the whales for good, and the tour companies knew that the whales were their biggest drawing card. Every boatload of tourists out of Seward was delighted with the sea otters, puffins, and calving glaciers, but every boatload desired more than anything to spot a whale.

Still, some boaters continued to chase the whales down. Twice this summer, a tour boat had collided with a whale—humpbacks both times. Too many boats were getting too close and crowding the whales. Encounters like this were a real danger: they could be fatal for humans and whales alike.

"The whales in Resurrection Bay deal with boats on them fifteen hours a day in summer, and they're still here," Eva said. I was as amazed as she at their tolerance.

Working with orcas in Puget Sound, Washington, Mercedes said she had seen as many as a hundred boats watching a single pod. The three Puget Sound resident pods had been in decline since 1995, and some suspected a primary culprit was the stress and interference in feeding and communication caused by whale-watchers. Recalling how quiet Prince William Sound had been in comparison with Resurrection Bay, I wondered why these whales stayed at all.

While we talked to the sailboaters, a tour boat did precisely as Craig advised. The AJ pod, porpoising steadily through the waves, swam right up to their boat; dozens of people lined the railings of the deck; the whales spouted and arced, some moving around the bow, some around the stern, as the cluster of people followed them, crossing to the other side of the deck.

"We've got people with tears running off their cheeks here," the captain radioed to us.

I was happy for the boatload of people, happy that they had seen whales, happy that Craig's method helped lessen the disturbance to the whales. But I couldn't help but wonder at what point even this would be too much: too many boats waiting for whales to approach, the constant pressure forcing the whales to leave.

I felt a wave of unease wash through me. I was here because I wanted to see these whales, too. I would have been disappointed if we hadn't. Riding with these researchers who were doing work that might help the whales did not exempt me from being part of the problem.

Perhaps there was some benefit to the whales in the long run—these people with tears running down their cheeks might return home feeling compelled to advocate for the whales' protection. But if we all had to see killer whales before we agreed to protect them, then there wouldn't be any whales to protect. There were simply too many of us.

I had never seen a whale in the wild until I came to Alaska. My knowledge of whales came through stories, and in the story of whales and humans was my first lesson in the loss of the wild. As a child I knew whales the way I knew dinosaurs—big, mysterious creatures who lived in a different world than I, and were, for the most part, "disappeared." The story of whale hunting was the first story I'd heard about people changing the world for the worse in a way that could be ameliorated only by restraint—by leaving the whales alone.

On a TV show about Keiko, the killer whale saved from a substandard aquarium in Mexico, wildlife biologists had debated the value of captive animals in winning people over to wildlife conservation. A spokesperson for the Humane Society had said aquariums were nothing more than "our little entertainment dens."

Maybe she was right. Animals in captivity don't behave the same way that they do in the wild; we have been given a false picture of them, which might be worse than no picture at all. Our imagination could bring more sustained connection—imagination fed through story and experience. Knowing one place well, one species well, could transfer to another.

As a child, I loved the book *The Story about Ping,* a duck who lived on the "beautiful yellow water of the Yangtze River." I wanted to be the little boy who lifted the basket and freed Ping from his parents' plans to cook Ping for dinner. I wanted to slip the metal rings off the necks of the cormorants who couldn't swallow the fish they caught. And yes, I wanted to float upon that beautiful yellow water myself one day.

When I saw ducks while playing along the French Broad River that flowed through my hometown of Asheville, North Carolina, I was reminded of Ping. Now, every time I saw cormorants here in Alaska, I remembered those cruel rings. Though I'd never been to the Yangtze, or even to China, my attachment to that river was strong.

When I learned that the Chinese government was damming the river, flooding the stunning canyons of the Three Gorges, it hit me like a lead pipe to the gut.

I started reading that book to my son when he was two years old. When he was in third grade, he became so upset that the State of Alaska was considering, once again, killing wolves in the name of predator control that he put together a petition. He gathered signatures from more than one hundred Alaska children, representing towns from Fairbanks to King Salmon, all asking that we reduce sport hunting of moose and caribou rather than sacrifice wolves. He sent it to the governor of Alaska. That Governor Knowles never acknowledged it didn't matter to Jamie; that the wolves stayed protected, at least until the next governor came into power, did. At the time, Jamie had never seen a wolf; he knew wolves through stories, through imagination.

Perhaps the tourists who saw the orcas in Resurrection Bay would transfer that experience. Perhaps they would help with a beach cleanup near their town, or lobby against logging in a national forest. Perhaps they would stop putting pesticides on their lawns to save the lives of the sparrows whose songs they heard each spring.

Our fascination with whales went way back; it was just the form that had changed. First we hunted them for oil; now we hunted them for pictures. These whales that had survived the long-liners' guns and the oil spill's poisons deserved our restraint.

@

All this boat activity out here was wearing, and I began to long for the still, quiet waters of southwestern Prince William Sound. I also knew I wouldn't see the AB pod here in Resurrection Bay. We saw, instead, the AJ pod, the AI pod, and then later the AN pod.

The AB pod, what was left of them, had been spotted in Prince William Sound the day before we had arrived. As we'd motored over, slowly in the heavy fishing boat, a seven-hour trip from Seward through the Gulf of Alaska to Prince William Sound, another researcher who studied humpbacks radioed to Craig that she'd spied the AB pod. But they had vanished by the time we arrived.

That was another reason I felt such an attachment to the AB pod: they seemed more loyal to Prince William Sound than other pods. So far, Craig and Eva had seen them only in the Sound, never in Resurrection Bay. I admired this fidelity to a place, especially considering the hard times that had visited them there.

In the mid-1980s, Craig began noticing that some of the AB pod whales had bullet wounds, some with holes through their dorsal fins, and several members were missing. He soon learned that fishermen were shooting them for stealing fish.

"They can't dive deep enough to get the fish, or to reach the lines on the bottom, so they wait until the fishermen are hauling in the gear," Craig said.

Craig and Rick studied ways to scare the orcas away from the longlines. They tried tricking the whales, pulling and then dropping the lines, working with another boat to lead the whales off track; they banged pipes together, used cracker shells, exploded seal bombs—nothing worked. Then they installed an "acoustic harassment device" that emitted a random foreign sound while they were picking the longlines. The first time, the whales began charging the boat, then stopped at the sound. All thirty orcas gathered together, lined up next to each other, touching fin to fin. Within two minutes, they moved en masse straight away from the noise. The next time Rick and Craig used this device, however, the AB pod ignored it altogether and swam straight for the fish.

Later, Rick had the underwater explosives used by some fishermen tested acoustically—a testing that led to the banning of these explosives as a deterrent. Fortunately, the problem was finally resolved by coincidence—because of potential overfishing, the black cod fishing season was reduced to just a few days a year. But not before nine members of the AB pod had been shot and eight had died. The pod recovered relatively quickly, rising to thirty-six whales by 1989, but those *X*s in the whale catalog were stark reminders of orcas killed before their time.

Then came the oil spill. On the third day after the tanker ran aground, Craig noticed some members of the AB pod surfacing right in the midst of a heavy slick of oil. He didn't locate them again until the next summer, and several whales were missing.

"When they come up to inhale, they expect to find air, not oil," said Craig. It was this inhalation of vapors and emulsified oil that most likely killed and injured the whales.

In subsequent years, Craig and Eva documented the disintegration of the AB pod. More and more members died; young were born, but more than half failed to survive the first year. The largest and most dominant maternal lineage within the pod went extinct. The two matriarchs died just before the oil spill, at least one from bullet wounds; two juveniles and two daughters died following the oil spill, and their orphan calves perished a few years later; the lone surviving male, AB3, with his dorsal fin collapsed, was seen swimming alone for years until he finally succumbed in 1996.

In 1993, a subgroup of AB split off and began swimming with the AJ pod. Like the high mortality rate, this was unprecedented. Orcas swim all the world's oceans, and people have studied them for decades in many parts of the world. But never before, anywhere, had anyone heard of part of a killer whale pod splitting off to join another pod. When one pod grows very large, sometimes it will split into two, but never had anyone observed some members abandoning one pod to join another. Each pod is a tightly knit extended family; each has a unique set of vocalizations, like a dialect. Joining another pod means learning another language; it means leaving the family.

A few months earlier, I had flown back to Ohio for three days to attend my grandmother's funeral. My father's mother, she had been born and raised in the mountains of West Virginia. As soon as she and my grandfather married, they hightailed it to Ohio, and she never went back.

"All those steep mountains and winding roads," she told me once. "I don't miss them at all."

After moving to Alaska, I traveled to North Carolina at least once a year to visit my family. Each time I drove between my parents' home in Asheville, my sister's home in Greensboro, and my friends' homes in Chapel Hill, the spaces in between the cities seemed to shrink, forests and pastures less frequent than buildings, billboards, and cloverleafs. I still loved the great variety of deciduous trees, the wide sandy beaches of the coast, the warm sun high in the sky. But each time I could feel the pull of North Carolina weaken and the pull of Alaska strengthen.

No one in my family had ever come to visit me in Alaska. Not my parents, none of my five brothers and sisters. For years, I had invited them, enticed them with photographs and stories.

"Come see me," I'd say to my brother Tom, who loved to fish, "and you can catch a hundred-pound halibut, a twenty-pound salmon." When my sister Angela was thinking of a trip to Switzerland, I told her, come see the alpine of Alaska instead. After a while, I stopped asking.

In Alaska, I was trying to re-create family, but it was hard. Jamie was with his dad nearly half the time; Rick was traveling for work, sometimes for months at a time. I had gotten a dog, a Siberian husky, just to have someone that stayed with me every night. But I'd chosen the wrong breed: Keira ran off whenever she got the chance—jumping out a half-opened car window, worrying a cable in half, bolting on nearly every walk—so I gave her two middle names: Vagabondo and Houdini.

©

The whales we finally found near Seward—the AJ pod—were those that some AB members had begun swimming with. My hopes of seeing some of the AB pod rose, even though the whales had never been seen outside the Sound.

A week earlier, Eva and Mercedes had seen the AB and AJ pods together in Bainbridge Passage, at the southwestern edge of the Sound. They had taken a smaller boat, the *Whale II,* over to see a tidewater glacier. On the way back, they had spotted the AJs and both groups of the ABs, essentially reuniting the AB pod.

"It was so nice to see them all in the Sound again," said Eva. "But I still don't think all of AB pod will get back together for good."

Today, though, there wasn't an AB whale in sight. The AJs, in several tight groups, were barely moving—group resting again. They surfaced in the same rhythm, porpoising forward but not traveling any distance, just keeping up with the tide.

"That looks like Zaikof," said Eva, pointing to a large male who suddenly moved over to the group nearest our boat. Zaikof was the name given to AB35, a male member of the AB subgroup that traveled with AJ. My breath caught.

"No, that's AJ4," Eva said, lowering her binoculars. I let out a long breath and loosened my grip on the railing.

Both Eva and Craig quickly and easily identified most of the whales they saw, as if they were picking out friends in a crowd. This came from years of observing them, being with them. This came from learning the whales' natural markings, not using tags or radios. This came from patience.

We were now in the skiff, moving beside the whales. Eva steered, and Craig took photos of dorsal fins. He kept trying to photograph certain whales, but AJ4, who was closest to the boat, kept getting in the way.

"He's less shy; the shy ones hide behind him," said Craig. Sure enough, every time they all surfaced together, AJ4 was right in front, blocking our view of the others.

We were within twenty feet of the whales, moving so close to them that, when they spouted, their spray coated my face. It felt like a mist, like a cleansing, an absolution. I was bathing in their expelled breaths when Craig noticed the tour boats had moved away, and he mentioned darting.

I got a rock gut and wanted to telepathically tell the whales, *Flee!* I felt chills over my body as if I had a fever. Hearing him, my mind raced, then stopped dead on the idea of whale hunting, an aversion so old it felt embedded deep within my skin.

Craig got out his gun. This, the body of a .22 rifle, was what he used to dart the whales for biopsies. Because it looked so much like a gun, he did this work only when other boats weren't around. He didn't want tourists thinking he was shooting whales.

That was what he was up to, but with a pneumatic rifle loaded with a pencil-sized dart that hit right below the dorsal and popped back out immediately. The dart had a sharp needle on one end and a bright-orange plastic plug on the other. After it hit the whale's back, it popped into the water, bobbing orange plug up. Then we motored by and retrieved it.

The dart collected a half-inch-long bit of whale skin and blubber in a hollow tube. This sample was sent off to a lab to determine the sex and age of the animal and document the levels of contaminants in their blubber. This DNA testing can also tell them who fathered a calf, so that they can discover the lineage of a pod more precisely.

After Craig darted a whale, the entire group stayed underwater longer. One whale did flinch when darted, its black shiny skin momentarily wrinkled. And another whale rolled over on her side, as if the force of the dart pushed her over, looking for an instant like she was playing dead.

But the side roll told Craig and Eva something—that AJ29 was a female. Born in 1989, she was still young enough that she could have been a male that hadn't sprouted yet. *Sprouting* is the term used to describe the male's dorsal fin elongating from the short, curved fin of juveniles and females to the long, straight fin of mature adult males.

Couldn't we just observe the whales, I thought, take their pictures, and in time learn the sex and age of every single whale? Couldn't we avoid darting and DNA tests? Couldn't we just wait for them to let us know?

"Yes, we could," Eva said quietly when I asked.

The next dart stayed in the whale a second longer, and all three of us gasped.

"It makes me shake every time," she said. It was a stunning thing, aiming a rifle at a whale, like unfortunate history repeating itself. I wondered if some encoding deep in the whales' memory was triggered when the gun was aimed.

Craig felt uncomfortable about the intrusiveness of the biopsy work, too, but he was convinced it was necessary for learning not just genetic histories, but also PCB and DDT levels in orcas. They were finding that transients had much higher levels than residents, and that females with calves had lower concentrations than males the same age, leading Craig to think that mothers passed the contaminants along through their milk. Such information might help them protect the whales. I suspected, though, that Eva regretted using this method of research on these whales they had watched so long. I wondered if this more intrusive method might affect the degree of intimacy they'd established with some of these pods, if the whales might become less willing to be in their presence.

Still, both Craig and Eva displayed far more empathy for the animals they studied than most researchers. One night in Stockdale Harbor at Montague Island, we came upon another boat of researchers who were studying harlequin ducks, trying to find out why these lovely sea ducks weren't recuperating from the oil spill.

I'd heard about their research project and the "harlequin roundup," an approach that seemed absolutely contrary to helping the birds recover. How must it have felt to these already stressed birds, when they were molting and vulnerable, to be encircled by people in kayaks, captured in a net, stuck with one needle to be drugged and another for a blood sample, and then awakened with a pain in the belly from the radio transplant surgery? Especially since most of those implanted with radios had subsequently died? Why take an animal that was already suffering and add to that suffering?

On the boat, which was much larger than ours, were several young, ruddy-faced researchers. They greeted us, and asked if we had seen any harlequins.

"No," Eva said quickly.

"Well, let us know if you do. We're having trouble finding them right now."

We had all agreed, as we approached their boat, not to tell them that we had just seen harlequins a few hours before off Channel Island, a flock of them floating serenely in the aqua sea.

"I don't understand the logic behind this kind of science," I said to Craig. "It seems unethical."

"Well," he said with a dry laugh, "ethics in research right now just means not stealing another's work."

At the beginning of his career, Craig had made a decision to avoid intrusive methods whenever possible. He had helped with a humpback whale study in the Sound in which they tagged the whales with brightly colored streamers attached by a metal flange shot from a crossbow. Once, after hours of chasing down three whales, they'd given up. Craig paddled over near the resting whales in a kayak. The whales breathed heavily, obviously exhausted. When they began to move again, one turned its tail to avoid hitting his kayak, demonstrating much more consideration for him, he noted, than he had for them.

Both Craig and Eva had told me stories of attempts to radio-tag whales. Not content with what they could learn through simple observation, some biologists had tried bolting radios on the whales' fins; inevitably, the whales had rubbed them off. One researcher had been particularly persistent. Once, he tried using a remote-control model helicopter to carry a radio fitted with metal clamps that, upon contact, would dig into the whale's blubber. On its first run, the helicopter had flown up, over the water, over the back of a whale, and back to shore, over the heads of the assembled scientists, and then over a nearby car, where it landed, clamping the radio onto the roof, ripping metal.

We had followed the AJ pod the length of Resurrection Bay now, around Fox Island, past a fleet of sportfishing boats. The Silver Salmon Derby had begun in Resurrection Bay, and what brought the sportfishermen here also brought the whales.

Everyone on board the *Lucky Star* had been talking about why the resident pods now frequented Resurrection Bay more than Prince William Sound. They had two theories, and both were probably correct. Resurrection Bay had been stocked with silver, or coho, salmon for the past five or so years. In Prince William Sound, all the hatcheries stocked pink, or humpy, salmon because they returned in just two years. But in all the scale sampling of orca feedings that Craig and others had done, they had never found pink salmon scales, only silvers. Like humans, orcas seemed to prefer the taste of the latter. Even in a school of pinks, orcas had been seen diving in and coming up with a wayward silver.

The other theory about why the whales now preferred Resurrection Bay was sobering: the AB pod's demise may have affected the other pods. Researchers used to see large aggregations of more than one hundred whales—superpods, they're called—every September in southwestern Prince William Sound, right where we had been those first four days in the Sound. A postseason party for socializing and mating, Rick called it. But no more. The AB pod was such a large and social pod that it may have drawn other whales to that area. Now that the AB pod had splintered, the other pods were dispersing—a ripple effect, a way in which what happened to one family had repercussions in other families.

Craig and Eva had also noticed that all the pods acted differently in Resurrection Bay than in the Sound. They spent more time feeding and resting, and very little time socializing. As if Resurrection Bay wasn't as much fun as the Sound, what with all the human activity. As if they were just putting up with us.

۞

Who was my family? The people I'd grown up with, who had never experienced the place I loved? Where was my loyalty, my taproot from which I drew life?

Like an oak, maybe, my roots had spread out, to reach from North Carolina to Alaska, but they had changed direction so that Alaska, Prince William Sound, held the root hairs, the fragile tips at the drip line that do the work of gathering nutrients and sending them through years of root and limb and trunk growth, send them on up to the crown, where again the smallest, most fragile, yet most vital part of the tree grows.

I could say clearly now what I'd known all my life: my family was more than human. It had always consisted as much of the wild places and lives around me as of the humans I'd known. In countless times of need, it had been mountains, or a particular curve in a stream, or the song of the golden-crowned sparrow that had held me within a solid embrace.

Several years after Andy and I separated, when Jamie was old enough to use language to describe his emotional life, I had asked him: How is it for you? What he said still made my throat constrict, my eyes well up with tears.

"When I'm with you, I feel like I don't have a dad. And when I'm with Dad, I feel like I don't have a mom."

It had taken days for me to find words to respond. How could I, other than to hold him so tightly that he would feel my embrace for days and days, feel the weight of my arms around his body, the warmth of my skin, the brush of my lips on his cheek, all through his time with his dad?

So, I told him a story. Just as I created stories for him every night at bedtime, I filled the dark with story. I told him about a hike I'd made alone in the red rock land of southern Utah; how I'd come upon a stream slicing through a sandstone canyon; how a hawk had alighted on a tree's broken limb, facing me; how I had stood on the ledge and held my arms out; how I'd felt it all embrace me, not just this hawk, this stream, this canyon, but the whole world. How I'd learned then what I wanted him to know now: we are all, each of us, every moment of our lives, embraced by the world. And every touch we receive is a manifestation of that everlasting embrace.

So even when you are not with me, Jamie, even when you are with your dad and you feel like you don't have a mom, just look at a tree outside your window, and see its arms reaching to you, holding you, and they are my arms around you, holding you, and they are the world's arms around you, holding you.

෴

The AJ pod began to head out of Resurrection Bay. We had followed them for a couple of hours, darting five and taking photos of most of them. They reached Cape Resurrection and began moving fast, swimming off toward Prince William Sound. I wanted to follow them, to head for the Sound myself, but I also wanted to leave them be.

We returned to Resurrection Bay and found the AN pod; for the rest of that day we kept up with them, listening and photographing and darting. The next day, we traveled west to Aialik Bay in Kenai Fjords National Park and tracked down another pod, but they were elusive, and we saw little of them.

As we played hide-and-seek with this pod, whales that clearly wanted to be left alone, I couldn't help but recall the stories I'd heard of the AB pod's friendliness. They would sometimes accompany the boat, Craig and Eva had told me, seeming to know the sound of the *Lucky Star*'s motor and seeking out their company. Some would even swim up and put their noses on the bow. Those that did, the ones that were most friendly, were now dead.

We humans had so many conflicting desires. We wanted to see the whales as predictably as we could in zoos, but without walls; we wanted to know everything about them, and we wanted to know now; and we wanted them wild and healthy, full of the mystery and grace that were the very pulls of our desire. But we were impatient and insatiable, unwilling to place limits on ourselves and mistrustful of our own imaginations, imaginations trapped like animals in cages.

The AB pod had trusted us. Once.

On my last day aboard the *Lucky Star,* Eva told me about a new pod they had seen one time, and then only briefly, a month earlier. The whales were out in the Gulf of Alaska, moving quickly, and the *Lucky Star* couldn't get close enough for dorsal fin photographs. But from what they could see, it was a strong, sleek, and healthy pod, showing few scratch marks. They had named it at once—the Magic Pod.

Bear

Picturesque Cove

Low tide, and the shallow cove is nearly drained. We haul the boats high on the beach, dragging their hulls across a field of dead and dying salmon. The acrid stench of decay fills the air as persistently as a cloud of mosquitoes on caribou. Their bodies, some crippled by death into grotesque forms, have faded from brilliant red into the same mottled gray as the mud that now holds them. A swarm of glaucous and herring gulls hop and flap, pecking at the worn-out bodies, snatching eyeballs from the still living, leaving them blind in their last moments.

Above the beach ryegrass, I walk the trail to the cabin. Beside it, the stream is clogged with fins like daggers carving water. At the sound of a splash, I turn. A few feet away stands a black bear, salmon in mouth. It lifts a shaggy head toward me, the small round ears twitching, then it gallops, bowlegged, out of the stream and up the near-vertical bank into woods.

I run to the cabin and latch the wooden door.

Later, I step out on the porch to hang wet clothes. A black bear in the stream, another on the bank, both freeze. I freeze, too. They stand no taller than I, but are thick with fur and muscle. If one were to lunge forward toward the deck, it would have me.

The bear in the stream breaks the trance: it swipes at the water with a wide paw, catches a writhing fish, dashes up the bank. The other enters the roiling water, high-stepping to avoid slipping on salmon.

Far into the late-summer evening, the bears fish the stream right beside the cabin. They must know I'm not a hunter. They must know that bear-hunting season arrives after the salmon are gone, when they fill their bellies with salmonberry and blueberry instead of fish, when they begin dreaming of long winter nights in dark dens.

scars we notice

1997–1999

SUMMER SOLSTICE is a potent time of year in Alaska. It's the longest day, twenty hours of light, but also the pinnacle—for on the other side, tomorrow and all the days to come for six months, we lose light each day. That hope all spring long, that gradual gain of daylight, each day longer, each morning brighter, turns to the realization that this is it: summer solstice. Better get out in it, we think, as anxiety creeps in, reminding us that tomorrow will be less bright, that time will now speed in the direction of seconds, minutes, hours lost, that summer solstice is also the beginning of the long descent into darkness.

So it was good that I was in Prince William Sound, on the water, in the summer air, leaves unfurling on the mountains all around us, snowmelt rushing down to saltwater, plankton blooming in the water beneath my kayak, gentle and quiet on the water. It was good that I was here on summer solstice, in full daylight for all the hours I could stay awake.

Three friends and I were paddling down Port Wells, southwest of the Harrison Lagoon Forest Service cabin we were inhabiting for the week. We stopped at a cobble beach where a rusted, canted truck served as an ugly trail marker to what was, at the beginning of the last century, the second-largest gold mine in western Prince William Sound. These were the ruins Andy had visited when he and I had stayed at the Harrison cabin with six-week-old Jamie. Now it was my turn.

The trail was an old road, steep but delineated with wooden

planks, and we hiked it quickly up to the mine site. The ruins were much more extensive than I had imagined. The skeleton of a building leaned against a steep rock ledge; two piles of crushed rock slid down between rotting boards; remnants of a house, complete with a chipped porcelain sink and rusted bedsprings and bits of broken plates and cups, splayed out next to it. Parts of unrecognizable appliances were scattered about in a muskeg meadow, where they were sinking, rusted brown, into the sphagnum moss as crowberry and bedstraw tendrils crept over them.

"It looks like they left in a hurry," said Sally.

"Yeah, like a forced evacuation," I said.

Following the road across the meadows and through forest, we came upon more scattered remains—a rusted oil drum, several wire hoops, a thick link chain looped around a dwarfed spruce. I tried to remove the chain, but over the course of nearly one hundred years, the tree had grown around it.

As we left I looked back at the detritus of human life: the rot, the rust, the tree subsuming the chain into its own flesh—it was like catching the land in the process of closing an old wound.

Evening, back at the cabin, we sat on the beach around a blazing campfire, Lisa's clear voice filling the air with song. I looked out over the water, still watching for what I had not yet seen: sea otters. Every other time I had been here, I'd seen a raft of sea otters offshore. I worried about what may have happened to them. Already so much I'd come to count upon was no longer here: the seal gliding in the lagoon at high tide, the orcas porpoising by, harbor seals sunning in front of glaciers. But the sea otters—their absence left me feeling heavy. My friends didn't miss these things; Gretchen had been to the Sound only once, last summer, and the others were here for the first time.

But even more than what was not there, I was disturbed by what was. All day every day, boats passed by in Port Wells. At least one large boat loomed on the horizon at all times, straight lines and sharp angles discordant. Engine noise always intruded, wherever we paddled, whatever time of day, a deep drone that reverberated across the water. I wondered what the din sounded like underwater, where the seals and orcas and halibut dwelled. My friends found the boats interesting; they looked at them through binoculars as if spotting a breaching whale.

When Andy and I had been to this area ten years earlier, we had seen only two other kayakers and the one tour boat that had plowed by, creating a chaos of noise and wake, causing nearby wildlife to duck and cover. Now such disturbance was constant. Now it was rare to not see at least one other boat—cruise ship, kayak, tour boat, or fishing vessel.

©

The nightmares started as soon as Andy and I bought our ten acres of land in Chapel Hill. Usually I dreamed that the hundreds of acres around it were cleared, and I'd wake up to find rows and rows of condominiums pressed up against the creek bank, towering over our little home.

We had sold the land when we divorced. I wanted to keep it, but couldn't afford it on my own. Rather than sell it to the highest bidder, though, we sold it to Peter, one of two people with whom we'd purchased and subdivided the original forty-five acres. Peter planned to keep it forested as part of a state small-forestry program; he promised to leave it be.

Right after the sale, I dreamed of the land again. Andy and I were out at the shed, sorting through the things we'd stored. The roof was rotting and letting in rain; plants grew from my old stuffed chair, vines climbed the bedposts. Andy and another man were grading the drive, and their truck kept hitting the roots and trunks of the big oaks, the sweet gum, the holly tree. Across the little stream, a boy rode a dirt bike; the smell of exhaust swept up the hill to me. Workers were building a massive power line down the middle of the stream, reshaping it, stirring up orange mud that spilled like blood over the running cedar. In my dream, the power line was for a new subdivision on the land that had been owned by Mr. Trice, a bricklayer whose family had been given title to the land by the owner of their slave ancestors. The half-mile-long drive to our place was now paved and littered with houses packed close, every one of them with a tightly trimmed lawn. But the land on the other side of our place, the Durhams' hundred acres, was still in forest. I started to plot how it could be protected, how it could be added to Peter's land and become one hundred and fifty acres of protected forest, an island big enough for some wildlife.

Some days I felt as if I was living the essence of those nightmares, where something I'd once known as whole was being dismantled. Not long ago, I'd run into a friend who had lived and fished the waters of lower Cook Inlet for more than twenty-five years. I confided to her my agony over the changes to Prince William Sound since the oil spill.

"It was the first place I'd ever seen where I felt as if I wasn't too late, that I'd found it before it had so deteriorated that it had become something else," I said. "But I think if I'd come to Prince William Sound now, I'd feel as if I was too late."

"Really?" she said.

My heart sank at the surprise in her voice.

There were few people whom I could talk to anymore about the oil spill, about its continuing effects, about the degradation occurring in the aftermath. Friends and family were weary of it; for them, the spill was long over, and they would look at me as if I were being indulgent, selfish, a look I recalled from elementary school that the nuns in the lunchroom used to communicate their disapproval when I failed to finish my meal: *How can you waste that food when there are children in India who are starving?*

I had thought it would feel good to be out with these three women who hadn't known me before the spill, before the divorce. I had thought it would allow me to step forward and reconcile myself to what was left, even find a way to embrace it. Instead, I was malcontent, as if I had brought my own rain cloud with me. I was saddened by the ways in which Prince William Sound was being diminished, I was still grieving the losses from the oil spill, and I was missing my son. Whenever I was without Jamie, I felt like I had a hole in my heart. The most enduring pain from my break with Andy arose from this: I had given up half of my time with my son. Half. The dark cloud carried that pain of missing my child, and that continual worry about how so much time away from his mother was affecting him. Rather than finding solace in the Sound, I was finding more to grieve, more to miss—and the hole in my heart just grew bigger.

Since the oil spill, paradoxically, Prince William Sound had become famous. Year by year, I watched the changes, watched as one kayak rental business in Whittier grew to four in Whittier, two in Valdez, one in Cordova; one Whittier day-tour boat became six, each of them carrying up to one hundred people; two water taxis to ferry campers and kayakers out and back became twelve. The U.S. Forest Service cabins that we used to drop in on if the weather was windy or wet began to fill up six months in advance, and reservations could be made only through a national reservation system located on the East Coast. The number of black bears killed by sport hunters each year doubled to as much as four hundred bears. Brown bear–hunting numbers skyrocketed as well. Bounded by sea and ice field, these animals had nowhere else to go. The list of species endangered, threatened, or at risk grew.

As did the pressures. Telecommunication towers on Perry Island were planned so that boaters could use their cell phones; increasing numbers of Jet Skiers roared unrestricted into estuaries; shoreline recreational subdivisions were announced for Passage Canal, Shotgun Cove, Poe Bay; permit requests flooded in for floating lodges, fuel docks, floating bars.

Beaches that once had no signs of human use were now littered campsites, little more than trampled bare spots among beach ryegrass and in slow-growing forests. At West Twin Bay on Perry Island, berry bushes and small spruce had been chopped down for kindling; three large blackened fire pits marred the beach and the forest; assorted pieces of plywood formed a privy among three trees, a wind block, and a railing to hang wet clothes.

At the Hobo Bay beach where my three friends and I stopped, I took a walk alone. When I turned from the beach into the forest, I came upon a mossy glade defiled with dozens of piles of toilet paper and human waste.

The damage to the Sound is incremental, it is cumulative, and it is often invisible. The damage to the orcas from PCB and DDT, to our own bodies from ingested and breathed automobile exhaust and chemicals, to our own hearts from crime and crowding and greed and power: all this is invisible and incremental, but it is also cumulative, often deadly. Even when it isn't entirely invisible, the incremental pace of degradation can make us blind to it. It is much harder to rally against tropical rain-forest clear-cutting—a slow but steady erasure of virgin forests and entire species—than against an oil spill—an immediate and vast disaster that kills thousands instantly.

Why do we call things disasters only when they happen swiftly? These incremental, cumulative damages are an irreversible erosion of the place. The permanent degradation of a place by increased access and overuse could, in the long run, be far worse than a catastrophic event like the oil spill. Chesapeake Bay, the Everglades, the California coast, the wild prairies—no oil spill, no fire, no bomb had desecrated these places. It happened with one more outboard cutting a wake, one more factory dumping waste, one more exotic plant taking hold, one more acre of wetland filled or forest cut or beach levy built for just one more house.

We accommodate these incremental damages, adjust ourselves to them, learn to ignore them, just as we adjust to the poisoning of our own air and water, to the noise pollution that deafens us. Scientists call this phenomenon "shifting baselines:" judging the amount of deterioration based on an already deteriorated environment. Our vision of a healthy, intact coral reef, for example, is now based on one that is already much less diverse, much less fecund, than what we considered a healthy coral reef twenty or fifty years ago. In Denali National Park, scientists installed stations to record the frequency and duration of sounds at different places, times of day, and season. They created a "soundscape" of the park for managing uses. But this data delineated an already degraded soundscape, one where flight seeing, snow machining, motorboating, and even the distant sounds of trains and trucks had already marred the place. How do we now define *quiet*?

We shift our very notion of reality. We name it, we study it, we acquiesce to it. We roll up the car windows in traffic, plug in headphones, build fences between our homes and our highways. We avert our eyes. How else can we get out of bed each morning?

Mine is a generation raised on the diminishment of the wild. Earth Day, the Endangered Species Act, the Wilderness Act, the Environmental Protection Agency, all began in our lifetimes. So did the daily extinction of entire species, acid rain, rampant rain-forest destruction, global warming. In my childhood state alone, as I grew into adulthood, black bears nearly vanished from the Great Smokies, red wolves became extinct in the wild, the forests of Mount Mitchell died from acid rain, and rivers like the French Broad and Haw became so polluted with industrial waste and runoff that we could no longer swim in them. Gone: it is a story line so common that it is woven into the fabric of our consciousness, all within only a few decades. But I could not, would not, allow the injustices and unnecessary losses to become unremarkable.

More than once, I wished I were my friends, simply enjoying this place, not haunted by memories of what was. I saw them reveling in the beauty, and I sat next to them under my black cloud. I didn't want to ruin their enjoyment of the place by telling them how things used to be, so mostly I kept silent. Why should I point out the constant engine noise or the dearth of sea otters?

But it frightened me that the damage was invisible to them. I was afraid that the degradation was happening so slowly that it wouldn't be noticed until it was too late, the way a frog lowered in cold water brought slowly to a boil never bothers to jump from the pot.

Still, what would my friends gain by knowing what had been lost, other than remorse and regret? Maybe people who had never seen the Sound before the oil spill ought to be allowed their delusions, even if temporary. Maybe I shouldn't come here anymore, dragging this dark cloud. Maybe I should just preserve in my memory the *locus amatus* I knew in my first few years here, and leave it at that.

But I could not abandon Prince William Sound any more than I could abandon my son. By love's alchemy, the divorce had strengthened my devotion to Jamie, just as the oil spill had strengthened my devotion to the Sound. I might have given up time with my child, but I would not give up parenting him and loving him absolutely unconditionally. Come what may, I am his mother, and I will love and protect him to the end of my days. Come what may, I will love and protect the Sound to the end of my days, as well. Even if I never see it again.

Should I move to the other end of the country, I would always be thinking of this place, wishing it well, wondering how it was doing. I would always be reminded of it by the place around me: oh, yes, that plant is related to the sundew that grows in the Sound's muskegs; that loon may well nest on a pond on Perry Island. There is a cemented bond, a current thick as blood between this place and me. There is no letting go or giving up. No end.

I stood on the ferry boat that runs from Hatteras Island to Ocracoke Island on North Carolina's Outer Banks. Across the water toward Ocracoke's southern point and the Atlantic Ocean beyond, a single kite tail looped in the ocean breeze. Then there were more, and they grew larger, twisting, expanding, contracting, like smoke curling from a chimney. They were birds moving toward us, long ribbons of black birds flying in from the sea. Through binoculars they looked like cormorants, but I wasn't sure. Did cormorants come here in winter, or did they just look like cormorants to me, because it was a bird I saw every summer in Prince William Sound?

A woman with binoculars around her neck walked by, threading her way through the cars on the ferry deck, and I asked her what kind of bird they were.

"Double-crested cormorants," she said. "They flock together by the hundreds this time of year."

I watched the undulating stream of cormorants advance. I felt good seeing them, knowing the same kind of bird that wintered in my childhood land, North Carolina, also nested in my chosen land, Alaska.

Jamie and I had spent two weeks visiting family and were taking a few days at the trip's end to travel the coast, the one part of North

Carolina where I could still imagine living. The rest of the state, the community of Chapel Hill, and the mountains around Asheville were now so congested, so void of undeveloped space, that they had left me feeling hemmed in and anxious. Out on the water, though, I relaxed.

Not that the coast had been spared. Last winter, Jamie and I had traveled north of Nags Head to Duck and Corolla, places that were, in my memory, small towns surrounded by dunes fixed by stands of sea oats and wind-sculpted forests sheltering wild ponies. What we saw, though, was resort after resort, huge hotels, condominium complexes, neon-green golf courses, a commercial vista without end. We had finally turned around at the Corolla lighthouse, a lovely old red-brick lighthouse that was dwarfed by the four-lane highway and clusters of new condominiums surrounding it.

This year we planned to travel in the other direction, toward the mostly undeveloped Cape Hatteras National Seashore, a string of sand islands that lies between Pamlico Sound and the Atlantic Ocean. Pamlico and Prince William, two sounds. They both are a haven for wildlife, protected waters along coasts that can often be harsh and storm-ridden. This shifting strip of sand called the Outer Banks blocks seas that have taken so many ships they've been called the "Graveyard of the Atlantic"; the islands of Montague and Hinchinbrook guard Prince William from the notoriously dangerous Gulf of Alaska. Both sounds contain all the nutrients of the wide ocean without enduring its roughness. Both hold fingers of water, small islands, river deltas, and marshlands, sanctuaries for nesting and rearing, and the food that comes in places where waters mix—the mixing of freshwater and saltwater at a river's mouth, the upwelling of warm over cold at a glacier's face. Both are places of commingling.

At Bodie Island, Jamie and I walked beyond the black-and-white-striped lighthouse that had stood sentinel to ships for a century, to a boardwalk that led out over the marsh. A few small birds sang to us from the tangle of bushes and cattails. We climbed steps to a wooden platform and looked out over the marsh, basking in sunlight as warm as a summer's day in Alaska.

The still waters were full of familiar birds. Clusters of white swans floated and dipped, upending to reach tender shoots beneath the water. They were tundra swans, the same bird that passed through Prince William Sound on their way to northern nesting grounds. Another kind of bird I saw all summer long was here in winter—Canada geese. A V-formation of them passed over our heads now; we heard the same cry that heralded the arrival of spring and the passing of summer in Alaska. Several passels of mallards and northern shovelers sliced the air in long curves over the water, flapped, and landed, flat feet sending up plumes of water. In the shallows near us, a pair of mergansers paddled by silently, reminding me of the mother and chicks I saw near Nellie Juan Glacier my first summer in the Sound. The chicks had still been flightless, so mother and babies had flapped their wings furiously, a churning line of birds moving away from us across water dotted by icebergs.

Another pond held a flock of birds, all white except for necks stained reddish brown and beaks thick and yellow. They were snow geese, here for the winter, great flocks passing through Prince William Sound in spring on their way farther north. One of the first birds to return to Alaska, they were harbingers of summer, landing in newly thawed marshes days after the first tender greens had sprouted. Here they pulled at water plants that grew all winter long. They made short, soft honks and clucks and quacks, not the loud call of alarm or arrival or direction, but the low, steady sounds of conversation, like the hum of a roomful of people who know each other well.

I thought about the years I'd spent exploring the string of sounds along North Carolina's coast. Perhaps they had imprinted on me, allowing me to see Prince William Sound as home, though it was in many ways such a sweepingly different landscape. Perhaps my love for this place allowed some access, through the cormorants and snow geese and swans, to the glaciers and ice fields and deep, cold waters of Prince William Sound. Like the migrating birds, my affiliation had transferred, so that, like the birds, I had come to need them both.

As Jamie and I drove back inland, across the wide plains of eastern North Carolina, back over the lovely arch of the coastal bridges, I realized these roads did not seem intrusive. They were here long before I first saw this place; they seemed somehow appropriate. I liked them. When I crossed over them toward the coast, my skin tingled with the same anticipation I had felt on childhood trips to the beach.

But there was a time when they weren't here. There were people who remembered what North Carolina's Outer Banks were like before any roads connected them to the mainland. What did they feel when they crossed these massive bridges? What was erased when these roads were built? What would I never know about this string of barrier islands and inlets and sounds?

The scars we notice are those inflicted during our watch, but they are all still scars.

๑

Once, as I moved through the train cabin selling tickets on the Whittier Shuttle, I'd overheard two fishermen talking about the idea of building a road from Portage to Whittier. I heard about that idea again and again over the years, but these fishermen, like everyone else, referred to it in the same way they would talk about an undersea water pipeline to California or a road to Nome or any other Alaska pipe-dream project: laughing and shaking their heads.

"Can't be done," one of the fishermen said.

"Even if it could," said the other, "why do it?"

Now this pipe dream was becoming reality. The Whittier Shuttle was to be replaced by a road from Portage to Whittier. This $80 million construction project linked Whittier and western Prince William Sound to the state by road, putting it within a ninety-minute drive of two-thirds of the state's population and the major tourism hub, Anchorage. The State of Alaska expected that, within five years, the road would increase the number of people visiting the Sound from 100,000 to between 1.5 and 2 million a year. The vast majority of this fifteen- to twentyfold increase in the number of people in the Sound would be concentrated in the northwestern areas—Port Wells, College Fjord, Harriman Fjord, Culross Passage, Blackstone Bay, Port Nellie Juan.

This road project proceeded even though its destructive effects would, in the long run, far exceed the devastation wrought by the oil spill. Not so much the actual building of a road but all that would inevitably arrive in its wake, all that would inevitably fan out beyond it, into the Sound. Local, state, and federal representatives had gathered at a public workshop in Anchorage months before the road would open to share their plans for dealing with increased visitation; instead they had shared their fears.

"Pulling the stopper on the road tunnel will be like a flash flood," said one U.S. Environmental Protection Agency official. "No one sees it coming until it's too late."

Their lists of increased risks spanned public water safety, unprecedented levels of disturbance to wildlife, degraded air and water quality, and shoreline habitat destruction. They spoke as if they were preparing for a disaster like the oil spill, not a planned and consciously executed public transportation project.

"There will be lots of environmental insults occurring over a period of time that are out of control," the EPA official warned. "I'd just as soon Prince William Sound stay in the protection category rather than moving to the reclamation category."

The reason to build the road, according to the governor's office, was this: to improve access to Anchorage for the residents of Whittier. At three hundred residents, that was a per capita expenditure of $250,000.

The unofficial word was different. Time and again, Whittier residents, shop owners, charter operators, and private boaters told me the road was built for industrial tourism.

Returning from a rainy camping trip, waiting in Whittier for the next train, I had stopped at the Hobo Bay Trading Company for some hot coffee and a brownie. Babs sat at the picnic table next to the door, smoking a cigarette and talking to another woman, one hand absently stroking her dog's back. She jumped up as I approached and greeted me with her usual hearty hello. As we went inside and I placed my order, I asked her what she thought of the road. She turned to pull some fries out of the deep fryer, then turned back around with a dark look on her face. She answered in a voice loud enough to carry to the harbormaster's offices next door.

"It's not being done *for* us," she said, "it's being done *to* us."

As Babs decried the unnecessary cost of the road, the undue pressures it would place on her community, I thought of all the hidden costs. I knew then that we'd pay for this road in ways we'd never know. If we knew, we'd think the price too high. But we would feel it, even if we couldn't name it, as a gaping hole in the soul, a hole with no name, a loss like that of a miscarried child or the untimely death of a lover. So difficult, so impossible, for anyone outside the grief to understand. And so very, very real.

As the train pulled out of Whittier, the train agent told me, "Tourists think we're crazy to have a road instead of this train. The Whittier Shuttle is the last train in the country that carries both vehicles and passengers. This is the end of an era."

I looked out the window for one particular rock outcrop in between the tracks and the water, the one where Peter the brakeman had lived that first summer. He had pitched his wall tent among a cluster of alders and wildflowers, right next to the water but perfectly secluded from the train, a hidden island of a home. But now I saw that the alders, and the rock outcrop, were gone, ground down into gravel bits for the road.

Dan, who'd been conductor the summer I sold tickets on the train, had been able to live in Whittier, selling fish and chips at Swiftwater Seafood, for only a year; the throat cancer that had cost him his train job quickly cost him his life. I wondered what he would have thought of the road to Whittier.

I loved riding that train, loved leaving the car behind and letting someone else drive, loved being lulled by the clackety-clack of iron wheels over iron rails, loved to watch waterfalls dancing down mountains on one side and the cloud-blue waters of Portage Creek swirl by on the other. The sounds of a train were older than cars and planes; the rhythm was like a heartbeat.

The first time Jamie rode the train, I held him on my lap in a passenger car. He fell asleep within minutes, even faster than he did on car rides. When he was a toddler, he liked to wander the aisles, stopping to look out the windows. In the tunnels, he would stare out the window, as if he was wondering where the world had gone.

When he grew up, what would Jamie remember of the train? He might not look at the road as a scar; he might consider it a beautiful drive. My friend Lois, who had put me in a kayak my first summer on the Sound, reminisced with me about our adventures in the Sound. We talked about the road and increasing popularity of the Sound, of the crowds and the changes looming on the horizon.

"Oh, well, I guess we had our fun," she said with her ever bright smile as her two toddlers clung to her legs. I was surprised at her easy acceptance; I wondered if it was the truth.

Raised on diminishment, we expect it. We resign ourselves to it. But what do we raise our children on? What of Lois's two children, growing up at the edge of this wild yet debased place? What could I give my son that would stem the tide? How could I raise him on reverence and love and connection with the wild, both for his sake and for that of the wild?

On our trip to Green Island in 1997, I had shown Jamie how the incoming tide changed the lay of the land, moment by moment. What had been a rock outcropping became an islet; what had been a tidepool was swallowed completely by the dogged force of the incoming tide.

The tide has cycles, dependable cycles of movement in and out. The inexorable change occurring in Prince William Sound might be cyclic in geologic time, but certainly not in my lifetime, or that of my son. Years before Jamie was born, I had talked with a biologist who had devoted his career to finding out why Steller sea lions were disappearing. He felt sad, he told me, that his one-year-old daughter would probably never get to see them in the wild. It was a presumption of loss that stunned me.

I wondered what Jamie would know of Prince William Sound when he was grown. I hoped that he would see sea lions draped across rocks at water's edge, the brown velvet of their fur glinting in sunlight; stand on a beach and hear only the waves and the distant keening of a pair of loons; drift before the face of a tidewater glacier, waiting in silence for ice to crack and keel into the sea. But I didn't know.

He would hear stories from his parents, stories that might remind him how the place had diminished, stories that might make him feel about Prince William Sound the way I had felt about so many places I'd known—that he was too late.

6

I searched for a different way to view all this change—if not for my sake, then for my son's. Each year for my birthday, which falls on the spring equinox, I'd come to Portage Valley. Sometimes the ice was thick enough on the lake that I'd ski out over it, ski around icebergs cemented in for the winter, ski to Bear Valley. Now with the building of the road, I had stopped coming.

The road had scarred Bear Valley, so that the magic I used to feel as we came out of one tunnel into the wide valley was gone—the valley floor was now littered with huge mounds of gravel for roadbed, crushed under the weight of heavy machinery, their grumbling engines shattering the silence, dismantling the landscape rock by rock, the way a trapper dismantles a wolf piece by piece, until all that is left is a charnel heap.

At the Begich-Boggs Visitor Center in Portage, I once stood before its enormous plate-glass windows overlooking Portage Lake. When the center was built in 1986, Portage Glacier's face was visible from this window, and the lake was jammed with the ever changing sculptures of icebergs. In less than twenty years, the glacier had receded by a mile, so that all I saw now was water stretching back to the mountains, and a few small bergs floating harmlessly along one edge.

I left the center and walked down to the lake's edge. Ten thousand years ago, this valley, and the entire Sound, was covered by ice so thick that only the highest peaks protruded. I stood by the lake, only the faint crackle and drip of an iceberg just offshore breaking the silence, and tried to imagine a sheet of ice pouring toward all this, the visitor center, the road, me.

This was the natural way of glaciers—they retreated and they advanced. Only one hundred years ago, the two-mile-wide entrance to Harriman Fjord had been nearly blocked by a glacier, one that poured from Barry Arm. When the Harriman Expedition of 1899 had come upon this face of ice, Harriman asked his most intrepid passenger, John Muir, if they should attempt the narrow opening between glacier face and rocky headland. Muir replied that Harriman might not want to risk his ship, but Muir wished to take a rowboat and go in himself. Instead, Harriman ordered the captain to take the ship through—and they were perhaps the first humans to lay eyes upon this twelve-mile-long glacier-streaked fjord. Less than one hundred years later, that glacier had disappeared.

Now, six billion tons of carbon dioxide spewed into the earth's atmosphere every year, causing an unprecedented glacial retreat. In just the past couple of decades, so much had changed. All in a time span far less than the life of a human, much less than the life span of a spruce tree, or the age of a mountain. With change so swift, and so devastating, sources of hope were hard to find. I recalled hearing about a rockfish recently caught in southeast Alaska: using the otolith, or ear bone, biologists had determined the fish was 204 years old. Such news heartened me, gave me comfort in the long view. But then again, the fish had been caught, its life ended. So many other lives were ending prematurely in this new age of extinction that we had brought on.

Still, there were forces greater than ours, and time was one. Despite this catastrophic global warming and glacial retreat, a few glaciers in the Sound continued to advance, reminding us of the complex cycles upon cycles unfolding in this world we inhabited. The human perspective was limited; a larger life force persisted, one in which mountains rose and fell, ice advanced and retreated, ancient forests marched through the natural phases of succession. When would the tide turn, when would the glaciers advance again? How long would it be before all this was undone?

In the chill of a sudden glacial breeze sweeping across Portage Lake, I could see it: the glacier advancing, relentlessly, forcefully, charging across the lake, encasing the lake in ice, crushing the roads and buildings in its path, pavement cracking and sinking and disappearing under the weight of ice, the glacier advancing into Bear Valley, down Portage Valley—leaving only the crystal silence of an ice-filled valley.

Glaciers advance and recede; change, as they say, is the only constant.

©

I dreamed I was on the Sound. Standing on the water. In Port Wells. Some of my friends were there in kayaks. They were trying to cheer me up, for I was full of sorrow about what was happening to this place. One was especially adamant that I get on with my life. Look how beautiful it is, she said over and over, pointing out every detail of the coast, the beaches and forests and meadows. Look, she said, at the sundew, how lovely and plentiful they are. And the porpoises, they're still here, and so are the sea stars. She listed every plant and animal, every lovely feature. The glaciers, they were not harmed at all, she said, her voice rising in frustration. I just stood, looking away from them and down Port Wells to College Fjord. She pointed to two killer whales passing by, proof that all was well, but we saw things differently. Yes, two orcas, yes, I said, but where are the rest? Maybe it's just you, she said, exasperated, maybe you like being sad. Maybe you're confusing the oil spill with your own personal tragedies so you won't be alone. Maybe you just won't let go of it because you're getting something out of it. I think, she said, you've blown this thing way out of proportion. The road, she went on, will just make it easier and cheaper to get out here, and it's a pretty nice-looking road so far, too. I walked off from them, their voices fading into the waves. I walked toward Harrison Lagoon, Barry Arm, Harriman Fjord. I began to move, slowly, a kind of dance, a series of movements like a bird stretching wings, a cat stretching legs, holding each pose a few seconds. The movement came to me as I did it. My friends were far behind me, specks on the horizon, watching, and they called it a prayer dance. I did it again and again, something in it older than our knowledge of this place. I danced on: elaborate moves, rolls and flips and bends, head, arms, back, and legs. I knew it should be difficult, but it was as easy as pouring water over my head; it flowed like rivulets down and through me. Jamie moved past them, too, and he was so small, the size of a doll. He began dancing with me, no words, just mirrored actions. He followed me, every step.

@

After four days on Port Wells with my friends, Andy brought Jamie out and left him with me. The black cloud lightened. For those two days with Jamie, time was suspended. I felt as if he would now be with me always, as if the Sound was now returned to its prespill self, the road only something people joked about, the wildlife still more abundant than humans. For those two days, I lived in the moment, no worry over the future or despair over the past. I stood on the beach with Jamie in sunshine, skipping stones and watching the ripples emanate infinitely outward. As the ripples reached the water's edge, I thought: *This could be enough.*

There were times when I thought I knew how to be a mother, and there were times when I thought I knew how to care for Prince William Sound. Other times I felt adrift, or confused, or outright incompetent. I tried to remember that the best I could give both was deliberate love. A love that knew no bounds—not of time, or of distance, or of change. Boundless love.

Perhaps what I was learning was this: loving a place unconditionally required a willingness not just to fight for its protection, but also to change along with it. That was the harder part. I knew the Whittier road wasn't right, and I had fought it as long as I could. Now it was time to accept the change. The road would be—at least for a while, until the next earthquake, or the next ice age. The oil spill had happened, and its effects would continue. The Sound would get only more crowded, developed, popular.

I had been naive to think that the Prince William Sound I first knew would remain, that this incredible place would not be discovered by others. Mine was the naïveté of the young, who don't know the intransigence and inevitability of change any more than they comprehend mortality. I couldn't hold back change, but I could refuse to avert my eyes to both the beauty and the horror. And I could believe that small acts might carry great weight, that their effects rippled outward long after we were gone.

On a trip to Harriman Fjord a month before the road opened, my friends and I came upon a fish seemingly trying to beach itself as the tide receded. It was a bottom fish, with a broad head and large shining eyes. I had never in my life seen such enormous pools of silver eyes on a fish. We watched it flounder upon the rocks and discussed what to do. One friend wanted to leave it alone, saying we should let nature take its course. But it was happening right before us; we were, however unwittingly, a part of what was taking place. I knelt beside the water's edge and gently cupped the long, shiny body between both my hands, lifted it up, and tossed the fish out into deeper waters. Seconds later, the fish was back at the shore, in danger of beaching again. I bent beside it. As I placed both hands on its slick skin, those silver pools looked straight at me. For a moment I was lost in their depths. Then I picked it up and flung it out once more. And once more it returned, this time remaining in shallow water just offshore. It swam up near to me. And I stood gazing at this fish, my palms adorned with translucent scales that caught the evening light.

Hummingbird

Harrison Lagoon

Jamie and I wander back into the cabin after a session of skipping stones on the beach. Inside we find two rufous hummingbirds, frantically bumping into the window. We had left the door open—and now they want out, to that feeder filled with red nectar they see just on the other side of invisible glass. We try to shoo them out, waving our arms, Jamie waving his hat. We are all four of us—mother, child, two hummingbirds—a frantic blur, and then Jamie and I stop and approach them ever so slowly. They are shimmering red and green bundles of energy, fluttering and bouncing off the window like rays of color emanating from a sunlit prism. I reach out my hands, a slow glide, encircle and catch one. It's a tiny thing in my hand, silky and light. It's as if I'm holding only air, except for a soft, quick pulsing, the hummingbird's breath. I coo softly to it, "Everything is fine. It's OK," was I walk to the door, out the door, and then open my hands and watch as it rockets out, landing on a blueberry branch. Jamie says, "Let me catch the other one," and I do, helping him cup his hands around it. I watch my boy's face as he feels that finite pulse, watch him as he releases it back into the infinite blue.

paradise falls

1999

EYES CLOSED, all I heard was the sound of icebergs aground on the black sand around me—icebergs melting, cracking, and dripping. Sculptures of ice the size of humans calved from one of three glaciers that, if I were to open my eyes and prop myself up, I could see: Cascade, Barry, and Coxe. I opened my eyes. Behind me, Jamie was scooping up handfuls of the fine sand and spreading them onto crystalline ice. Before me, Rick was climbing up a rock headland with his camera. He wanted this picture, and so did I; this day, the glimmering glaciers, the black sand, the sun. The three of us.

It was a long paddle to reach this beach along Barry Arm, one that was nearly in vain. After six hours of pulling against the tide, and just as we had come within sight of this beach, we first heard, and then saw, a large trimaran barreling into the arm. As it got closer, we could make out the running narrative croaking over its loudspeaker. Packed with a hundred or more people, this was a commercial boat on a day tour of the Sound. "Twenty-six-glacier cruise!" their advertisements promised, and all in four hours. My chest tightened at their appearance. Then a second tour boat plowed into view right behind the first. They both rumbled past us, loudspeakers blaring, and motored up closer to the glacier than where we now sat, quickly becoming the size of toy boats in a bath.

"Well, I guess we're just going to have to get used to that," said Rick.

"Yes," I said. "We'll just outlast them."

After less than a half hour, both boats turned and churned on past us again, heading farther into Harriman Fjord. Then another sound: the wonderful thunder of calving ice. On the far side of Barry Glacier, ice tumbled in a giant white cascade into the water, roared as it crashed into saltwater, sending long, deep swells radiating outward, reaching our shores several minutes later.

I hadn't planned on coming to the Sound this summer, but it was the one hundredth anniversary of the Harriman Expedition, whose members first discovered this glacier-rimmed fjord. One hundred years since John Muir had proclaimed Prince William Sound a "bright and spacious wonderland," since John Burroughs had called it an "enchanted circle." It was also ten years after the *Exxon Valdez* oil spill, and one year before the road to Whittier was scheduled to open. I felt I owed it to the place and to myself to make one last trip before the road changed everything. To remember, to honor, to savor.

Rick didn't want to come. He didn't want to go to the northwestern Sound because it had become so popular. Whatever trepidation he had, though, seemed to have been left behind in Whittier. We were in a part of the Sound he'd seldom seen, and every day he was up early, full of energy, wanting to explore every inch of this place. After the first two days, he told me, "Thanks for getting me out here. This is just what I needed."

What I needed, too. I'd been here many times; this was a chance to be in a place full of memories, of what I'd loved, what I'd lost and given up. All that was lost from the spill, all that was lost in my marriage, all that might soon be lost with the opening of the road. Still, I worried that I would dwell on the loss; I feared that the Sound had become for me a cause instead of a place, an issue instead of a beloved.

This time, though, this time, I was right here all the time, with my beloveds. We were all three of us right here and now, in the undeniable beauty of these mountains, forests, waters.

The first day had been rainy, recalling our trip to Pigot Bay the year before—six days of incessant rain. By midmorning, Jamie had been full of unfocused energy, bouncing around inside the cabin, repeating in a singsong voice, "I'm bored." Still unpacking, I'd felt agitation begin to flare.

Just in time, Rick had taken Jamie outside, shown him how to make a hatchet out of a stick, a stone, and a piece of rope, shown him how to strike it against a piece of driftwood at just the right angle to sever it. I'd found them out on the beach, looking like pirates with their shirts off and bandannas around their heads. By then they were making small offerings to the beach, creating works of art from gathered bits of seaweed, rocks, shells, and driftwood.

"Look at this one, Mom!" Jamie cried, grabbing my arm. "And this one!" He pointed to a curved piece of driftwood adorned with a spray of popweed, three white-and-brown stones, and a dark-blue mussel shell.

Now setting off from the black sand beach, we drifted together in a field of ice, listening to the crackling of ice melting in saltwater. From the back of the double kayak, Rick scooped up small chunks of ice with his paddle and passed them up to Jamie, who sucked on them contentedly. The glacier was now in him; he was now made of it. Would he have its power, its boldness, its thunder?

The paddle back took well into the evening, nearly eight hours. Rick told stories nearly the whole way: the adventure saga of an indomitable squirrel, until Jamie fell asleep, resting his head on his life jacket. He looked like a baby again, reminding me of the first time I brought him to Prince William Sound—six weeks old, huddled against my chest inside my coat; two years old, running on these beaches, grabbing fistfuls of sand; three years old, sitting in a meadow filled with wildflowers; five, climbing a rocky headland barefoot; six, in a single kayak on his own, paddling the edges of a quiet cove.

A shudder ran through my body, and I feared it would stay, caught like an echo bouncing off canyon walls. But it didn't, not this time. My memories would always be tinged with regret and sorrow over the ways in which I had failed to attend to my marriage with Andy. But now I could also feel joy in the living we did share, in the son we still shared. And there was the joy of this moment, that I now shared with two I loved.

I looked again at my sleeping son. Jamie's eyes were blue and his hair fair, like mine, but he had his father's hands, his feet. The same broad shoulders. For this I was glad.

Paddling, paddling, so tired the edges of my vision softened and took on the texture of a dream. All around were the stillness and quiet of a sun-soaked summer evening. I kept wanting to stop, to rest, but it was already so late. Rick's kayak continually outdistanced me, until finally he convinced me to let him tie my kayak to his. He had a way of putting his entire self into a thing, be it saving a forest or paddling us back to the cabin—this I loved, this I was learning from him: to not give up.

So we paddled on, quietly. The sun angled downward, bringing on the summer night. Before us, growing larger with each stroke, was the half-mile-long spit that guarded the entrance to the lagoon. The tide was out; now we'd have to paddle around the spit, me so exhausted I could hardly move my arms. But Rick kept us close to shore, hoping to cross it at a low spot. And we did, the faintest scrape of kayak on rock as mine cleared. Suddenly, a cloud of white kittiwakes arose from the shore and circled us in a crescendo of sound, as if applauding our return.

The next day we paddled across the lagoon to the waterfall, the tide nearly high. We beached the kayaks on a ledge near the falls and started climbing up beside them on a series of steep ledges. The falls, a series of three major drops, were so thunderous they drowned out our voices, so immense that they coated us in the fine, sweet spray within minutes. Just below the first drop, a great mist arose from a pool hidden behind a rock lip.

"It looks like dragon's breath," said Jamie. And so we named them Dragon's Breath Falls.

We followed the stream up into the woods, savoring its music and the lovely rain-dropped plants all around, the immense Sitka spruce towering over blueberry bushes and mosses. There was in this ancient forest a feeling of space and cover at the same time—the feeling of a virgin forest, that mix of mature canopy and understory harboring a diversity of life. The stream, strong and wide, rushed clear over boulders, changing shape with changing topography: fifteen feet wide and tumbling with abandon over a field of fist-sized rocks, narrowing and gushing in white plumes through larger boulders and a steep bank.

I found a flattened, concave moss-covered spot near the bank, and nearby a plate-sized pile of scat. A black bear lived here.

"I wish I were a bear, so I could live here," I told Jamie and Rick. They both stooped to poke at the scat with a stick. "In fact, I do want to live here. Let's."

A few steps away, a pile of lumber lay crumpled, rotting and overgrown with moss, as if someone had once tried to live here, enclosed in walls made of dead trees, walls that blocked the light and the sound of the water. No—that was not how I wanted to live here—I wanted to lie down in a bed of soft sphagnum moss, eat berries or shrews or salmon when I was hungry, drink from the crystal stream when thirsty. For one bright moment, a veil seemed to lift, and I knew I could.

But soon the walking became tricky, the understory more dense and filled with the prickly stems of devil's club. Our human legs stumbled more frequently.

"Time to turn back and find a trail," I said.

"No," Jamie pleaded. "Let's keep going this way. I want to keep going!"

Rick and I looked over his head and smiled at one another.

We walked a bit farther upstream, then backtracked and found a trail, following it up through muskeg meadows and up until we reconnected with the stream. Along the way, in muskeg and in forest, I counted sixteen different kinds of wildflowers, a pallet of color and shape, brilliant red, yellow, white, purple, orange, blue.

We followed the stream up to a cascading waterfall that seemed the stuff of fairy tales. Or were fairy tales the stuff of it? It flowed down in two steps of equal height, steep cliffs adorned with fragile ferns and saxifrage on either side, their slender stems rhythmically nodding in the spray. At its base spread a gravel bar full of white foamflowers, and among the boulders splashed with spray a blue-gray bird, a water ouzel, slipped in and out of the water. It gripped the rocks to walk under the roiling waters in search of food, then flitted over the surface, in and out of the spray, flying up to land on a niche in the cliffs among fern and saxifrage.

We skirted the cliffs and clambered up above the falls to a snowmelt lake at the edge of a cirque. Warmed by the climb, we peeled off our clothes and plunged into the lake, then jumped out quickly, our skin red and tingling. Above us treeless tundra, steep rock slopes, and jagged peaks stretched into blue sky. We drank handfuls of the icy water, then sat above the falls—which we named Paradise Falls—in a meadow of heather ringed by wind-pruned mountain hemlock, looking out over Port Wells, across the water glinting silver in the sunlight, to snow-jeweled mountains on the other side. All this beauty. Us in it. It in us.

On the way home, we made one more stop: the beach on the other side of the narrow lagoon entrance where an old wooden structure stood. Two pairs of logs five feet apart formed inverted Vs that held up a fifth log. I had always thought it was something that had been used by Natives at fish camps in late summer. But Rick said no; hunters used it to hang, butcher, and drain the blood from dead bears.

Knowing that, knowing what it had been used for, I wanted it gone, that which I'd seen from across the water for so long. We tugged and pulled and heaved, all three of us, until it was torn down, the logs separated, rope cut, all of it hidden by thick stands of beach ryegrass.

Later, I crouched on the beach, staring across to where the butchering rack had stood. I thought of those black bears in the deep woods, sleeping on sphagnum moss beside the perfect stream. I thought of their daily lives, padding through the forest, drinking from the stream, eating berries from the woods and salmon from the sea. I thought of the black bears I'd watched my first summer in the Sound, the way they'd fished beside the cabin in Picturesque Cove, the way they'd let us watch. The bear whose bed we had just this day found could be killed in a few short months by hunters who stayed at this very same cabin. Tearing down the rack felt like atonement.

For so many years, I had been waiting for some kind of ending, some return to the way things had once been. But now I saw that was not the way of this world; now I saw that I must go out into life with new eyes every day.

Perhaps that was what underlay the continual joy of this trip—the joy that came from living through loss, living through it even though it felt like trying to walk through solid rock, and finding myself still here, on this earth. Finding joy when I yielded to the river of life, to its waterfalls and eddies, trusting it, each moment more precious knowing how fragile it all was, how impermanent, how imperfect and scarred we all were. And how alive.

This was how it was for me now: I had this moment; this place, this son, this man. I thought of the yogi who broke his teacher's favorite cup. "Don't worry," said the teacher, "I have always drunk from that cup as if it was already broken." I was learning to love in the way that only feeling the loss of the beloved could provide. I was learning to pay attention, to take care, and to take nothing for granted. The lesson was inside me now, just as the waters of glaciers and snow were inside us.

Late that night, when the summer sky had dimmed, we three fell asleep listing the plants and animals we'd seen over the past few days. Our voices mixed, became a mantra:

> sea otter saxifrage chocolate lily melibe sculpin Steller's jay anemone false hellebore rufous hummingbird arctic tern hermit crab foamflower water ouzel mountain hemlock yellow violet ocher sea star sea lettuce Sitka spruce goatsbeard old-man's beard columbine glaucous gull currant blueberry salmonberry crowberry watermelon berry bog cranberry cloudberry sundew green orchid bog orchid purple iris devil's club limpet blue mussel purple aster scoter water hemlock common loon purple violet barnacle grebe yellow cedar bald eagle springtail popweed harbor seal sun star pushki ladies' tresses lupine kittiwake

epilogue

It's TRUE: you can't step in the same river twice. With each breath I take, water flows into and out of Prince William Sound, from the Gulf of Alaska, the Pacific Ocean. Swirls of plankton gyrate through the water column, nutrients moving in ocean songs, salmon and seabirds and whales following the siren call. Animals are born, move in and out and within the Sound, give birth, die. Plants emerge from thin soil, grow and blossom and fade, their seeds held aloft, set down on new ground. Ice cradled by mountains for thousands of years falls in thundering explosions. Glaciers recede and advance, trees rise and fall, solid granite lifts and subsides, shifts and erodes.

The earth moves. A post holding one side of our wooden gate levitated from the ground this spring, heaved up by frost over several weeks, like the dead arising from graves. As lengthening days warmed deeper down, the soil thawed and the post settled down just as slowly, several inches away from where it had started. The earth breathes.

You can't step in the same river twice—although I once believed I could. I believed that the pieces of my life I had chosen, those I held close to my heart, would, once chosen and held, remain the same. Call it the naïveté of youth, call it human delusion, human arrogance. I believed that if my idealized vision of life and the world was threatened, I need only bolt like a spooked horse; I need only run away. Now I stay.

Fifteen years after the *Exxon Valdez* hit Bligh Reef, single-hulled oil tankers, many of them nearing their twenty-five-year retirement age, still traverse the waters of the Sound daily. The United States uses more oil today than in 1989, and pressure intensifies to drill in Alaska's land and offshore waters. This oil we burn ever faster has brought on global warming, causing irrevocable damage to the land and sea and wildlife of Alaska.

Exxon continues to appeal the $5 billion punitive verdict from 1994, though their most recent first-quarter profits alone topped $7 billion. An additional $100 million available from Exxon has yet to be requested by the governments, though unanticipated damages abound in every corner of the spill region. Pacific herring have been downlisted from "recovering" to "not recovering"; the AT group of orcas—the only transients that regularly swim the Sound—is but a few members away from extinction; persistent organic pollutants are showing up in ominously high levels in orcas and other marine life. As always, numbers don't tell the whole story: at the Bird Treatment and Learning Center, in the midst of Anchorage's industrial district, a bald eagle known as "One Wing" still perches listlessly in his outdoor cage, ignoring a chunk of red salmon attracting flies in the midday sun—fifteen years after losing his wing in the oil spill.

The road to Whittier has now been open four years; visitation has increased threefold. A new upscale multistory hotel rises between railroad tracks and the water, and the number of cruise ships touring the northwestern fjords has grown by 50 percent. A small cruise ship, pushing farther into the recesses of the Sound's convoluted waterways, ran aground in June 2003 in Jackpot Bay. But the forests of the Sound are safe—for now—from clear-cutting.

A friend of mine, on board the twenty-six-glacier cruise of Harriman and College Fjords, was appalled when the captain announced in no uncertain terms that the Sound had completely recovered from the oil spill, that you could dig as deep as you wanted into any of the beaches and not find oil, that all the species had fully recovered.

"That's a lie," she whispered to two elderly women from Illinois who sat next to her.

"Really?" they said, pulling back in their seats.

March rolls around each year with the same demarcations: March 21, spring equinox and my birthday; March 24, the anniversary of the spill. I note my age and that of the oil spill—one with celebration, the other with remembrance tinged with sorrow. Now it's a private remembrance, the days of press conferences and group gatherings over. Perhaps some event will mark the twenty-fifth, but how many might attend? How many who were actually there? Who lived it?

Every major spill dredges up painful memories: Shetland, France, Spain. Mei Mei, whom I now rarely see, called the first week after the tanker *Prestige* sank off the coast of Spain in the fall of 2002. Old feelings had flooded her, enough that she felt the need to talk to me, to someone who would understand.

Despite the new road, I couldn't stay away from the Sound. When I drove it for the first time, I shuddered when we slid past Portage Lake and into Bear Valley on smooth blacktop, skirting the first tunnel entirely. But when we stopped at the overlook, I couldn't help admiring the view of the lake from a new angle. Was I accepting the change, or was I becoming complacent, allowing my own baseline to shift?

It is the second tunnel that cars now drive through, the dark canal of those early years widened and lit. At the end of the tunnel, with the waters of the Sound broadening before me, I awaited that sense of rebirth.

We parked in the new gravel parking lot that spreads toward Whittier Falls and slipped our kayaks into the water on a small beach just outside the expanding harbor. Rick and James and I followed the same route I'd taken my first time in the Sound—across Passage Canal to the kittiwake rookery, down the shoreline to the half-moon of beach backed by dead spruce forest. But we went farther, to Poe Bay to camp for the night, where we found a waterfall tumbling down slickrock, loud enough to drown out the sounds of passing boats. James, his eleven-year-old legs as strong and fast as mine, led the way up the series of smooth rocks angled along the course of falling water.

In the evening's waning light, an arctic tern perched at the top of a willow snag on a grass-covered islet. Another tern flew in, white wings slicing blue light, holding a single herring. He flitted above the first tern, offering the fish. The tern on the snag lifted her head and gently accepted the silver thread. They repeated this ritual again and again, and each time, she accepted.

Here, take this offering. My heart. May it be received into the slender life of an arctic tern. May it lift up into the air on feathered wings, dance across the water, glide out over Port Nellie Juan, Icy Bay, Dangerous Passage, along the shores of Perry and Knight and Montague Islands. Let it remain here long after I have gone.

acknowledgments

I AM INDEBTED to Dawn Marano, the editor who understood and helped peel back the layers, and to all associated with the University of Utah Press for their good work. My thanks to writing friends and editors who have provided insight, inspiration, and encouragement, especially Karin Dahl, Carol Hult, Gretchen Legler, John Murray, Jerry Mills, David Abrams, George Bryson, Tom Sexton, Ron Spatz, Tara Wreyford, Naomi Klouda, Melissa Walker, Terry Tempest Williams, Peggy Shumaker, Elizabeth Statmore, and all my Hedgebrook sisters. I am grateful to Hedgebrook and the Weymouth Center for residencies, the Alaska State Council on the Arts for a grant, and those who helped me be in the Sound, particularly Perry and Lois Salmonson, Kelley Weaverling, Craig Matkin, and Eva Saulitis. Special and enduring thanks to Andy, with whom I first experienced the Sound. Endless gratitude to Rick for his commitment to and knowledge of the Sound, and his unflagging support and belief in my work; to my son, James, for being a bright and steady source of light; and to Prince William Sound and its animals for sharing their lives with me.

Breinigsville, PA USA
25 October 2010
248029BV00001B/1/P

9 780803 230354